MATH KNOTS
Joy of learning....

AF779637

My Book

This book belongs to

Name: Anjali

Cover Design by :
Gowri Vemuri

First Edition :
April , 2021

Author :
Gowri Vemuri

Edited by :
Raksha Pothapragada
Ritvik Pothapragada

Questions: mathknots.help@gmail.com

NOTE : These tests are neither affiliated nor sponsored or endorsed by any organization.

Dedication

This book is dedicated to:

My Mom, who is my best critic, guide and supporter.

To what I am today, and what I am going to become tomorrow,

is all because of your blessings, unconditional affection and support.

This book is dedicated to the

strongest women of my life ,

my dearest mom

and

to all those moms in this universe.

G.V.

Visit www.a4ace.com

Also available more time based practice tests on subscription
Subscribe to Math-Knots you tube channel for concept videos
E-mail us: mathknots.help@gmail for more concept based videos with a proof of purchase.

The Iowa Algebra Aptitude Test (IAAT) was developed to help teachers and counselors make informed decisions about initial placement of students in the advanced mathematics curriculum. IAAT test provides a four-part profile of students that identifies specific areas of strength and weakness. This test is divided into the following 4 sub-tests and total test time is 40 minutes

1) Pre-Algebraic Number Skills and Concepts

2) Data Interpretation and Analysis

3) Mathematical Relationships

4) Mathematic Symbols

Different Counties conduct these tests in different grades. As an example, the following table lists the grades in which the IAAT test is conducted in Northern Virginia.

Fairfax County Public Schools (FCPS)	Grade 6
Loudoun County Public School (LCPS)	Grade 5
Prince William County Public Schools (PWCPS)	Grade 5

This test is a time-based test, for which speed and accuracy are important factors. Students are required to get 92% and above in the IAAT test and a pass advance in the SOL's/End of year assessments at school. To help kids excel in this test, **Math-Knots** published several books.

IAAT (IOWA Algebra Aptitude) Practice tests Vol 1	6 practice tests	https://www.math-knots.com/iaatpracticetests
IAAT (IOWA Algebra Aptitude) Practice tests Vol 2	8 practice tests	https://www.math-knots.com/iaatpracticetests
IOWA Algebra Aptitude Work Book - Vol 1	1000 practice questions	https://www.math-knots.com/iaatworkbook
IOWA Algebra Aptitude Work Book - Vol 2	1000 practice questions	https://www.math-knots.com/iaatworkbook

Topic	Pages
Preface	1 - 12
Notes	13 - 82
Fractions #1	83 - 88
Fractions #2	89 - 92
Fractions #3	93 - 96
Decimals #4	97 - 99
Decimal #5	100 - 102
Decimal #6	103 - 105
Integer Multiplication #7	106 - 108
Scientific Notation #8	109 - 113
GCF #9	114 - 116
GCF Monomials #10	117 - 119
LCM Numbers #11	120 - 122
LCM Monomials #12	123 - 125

Topic	Pages
Order of Operations #13	126 - 128
Verbal Expressions #14	129 - 131
Verbal Expression Equation #15	132 - 136
Monomials #16	137 - 139
Inequalities #17	140 - 143
Evaluate Expressions #18	144 - 146
Evaluate Expressions #19	147 - 149
Solve for X #20	150 - 153
Absolute Value #21	154 - 156
Absolute Value #22	157 - 161
Proportions #23	162 - 165
Proportions #24	166 - 169
Percent Discount #25	170 - 173

Topic	Pages
Percent MarkUp #26	174 - 177
Percent Tax #27	178 - 181
Percent Change #28	182 - 185
Radicals #29	186 - 189
Radicals #30	190 - 193
Radicals #31	194 - 197
Radicals #32	198 - 202
Radicals #33	203 - 206
Radicals #34	207 - 210
Triangles #35	211 - 212
Rectangle-Square #36	213 - 214
Parallelogram Trapezium #37	215 - 216
Circle - Area - Circumference #38	217 - 220

Topic	Pages
Volume-Sphere #39	221 - 224
Volume-Rectangle Square prisms #40	225 - 229
Volume - Cone - Cylinder #41	230 - 234
Additive Inverse #42	235 - 238
Multiplicative Inverse #43	239 - 242
Rational - Irrational Numbers #44	243 - 245
Exponents #45	246 - 249
Slope - 2 points #46	250 - 254
Slope intercept form #47	255 - 259
Slope Graph #48	260 - 263
Straight line slope #49	264 - 265
Parallel line slope #50	266 - 267
Perpendicular line slope #51	268 - 269

Missing angle #52	270 - 273
Missing angle #53	274 - 276
Reflection #54	277 - 280
Rotation #55	281 - 284
Translation #56	285 - 288
One step word problems #57	289 - 291
System of equations #58	292 - 295
System of equations #59	296 - 298
System of equations #60	299 - 301
Line graph #61	302 - 311
Scatter plot #62	312 - 316
Histogram #63	317 - 320
Stem leaf plot #64	321 - 326

Dot plot #65	327 - 331
Pie chart #66	332 - 342
Answer keys	343 - 364

Fractions Notes

Fractions are part of a whole. It is also an expression representing quotient of two quantities.

Example 1 : $\frac{2}{4}, \frac{1}{4}$

Fraction can also be represented as ratios.

Example 2 : 1 : 2 or $\frac{1}{2}$

3 : 7 of $\frac{3}{7}$

Adding simple fractions, follow the below steps :

#1 : Fractions with like denominators can be added by adding their numerators.

Example 3 : $\frac{3}{5} + \frac{1}{5} = \frac{3+1}{5} = \frac{4}{5}$

#3 : To add fractions with unlike denominators, convert the fractions to equivalent fractions with like denominators and follow #1. To convert them into equivalent fractions you can multiply the numerator and denominator with a common factor for each of the fractions to be added separately and then add the fractions.

Method 1 :

Example 4 : $\frac{2}{5} + \frac{1}{2}$

Step 1 : $\frac{2}{5} = \frac{2 \text{ X } 2}{5 \text{ X } 2} = \frac{4}{10}$

Step 2 : $\frac{1}{2} = \frac{1 \text{ X } 5}{2 \text{ X } 5} = \frac{5}{10}$

Step 3 : $\frac{2}{5} + \frac{1}{2} = \frac{4}{10} + \frac{5}{10} = \frac{4+5}{10} = \frac{9}{10}$

$\frac{2}{5} + \frac{1}{2} = \frac{9}{10}$

Another method to add fractions with unlike denominators is by finding the **Least Common Multiple** (LCM) of the denominators and then follow the steps as described below in same order.

Method 2 :

Example 5 : $\frac{5}{12} + \frac{1}{8}$

Step 1 : Find the LCM of the denominators 12 , 8

$$\begin{array}{r|l} 2 & 12\ ,\ 8 \\ \hline 2 & 6\ ,\ 4 \\ \hline 3 & 3\ ,\ 2 \\ \hline 2 & 1\ ,\ 2 \\ \hline & 1\ ,\ 1 \end{array}$$

LCM = 2 X 2 X 3 X 2 = 24

Step 2 : Converting $\frac{5}{12}$ into an equivalent fraction with the denominator as 24 (LCM).

Divide the LCM value with the number in the denominator of the fraction to obtain the common factor.

$$\begin{array}{r} 2 \\ 12 \overline{)\ 24} \\ -24 \\ \hline 0 \end{array}$$

Common factor obtained is 2

$$\frac{5}{12} = \frac{5 \text{ X } 2}{12 \text{ X} 2} = \frac{10}{24} \ ; \ \frac{5}{12} = \frac{10}{24}$$

Step 3 : Converting $\frac{1}{8}$ into an equivalent fraction with the denominator as 24 (LCM).

Divide the LCM value with the number in the denominator of the fraction to obtain the common factor.

$$\begin{array}{r} 3 \\ 8 \overline{)\ 24} \\ -24 \\ \hline 0 \end{array}$$

Common factor obtained is 3

$$\frac{1}{8} = \frac{1 \text{ X } 3}{8 \text{ X } 3} = \frac{3}{24} \ ; \ \frac{1}{8} = \frac{3}{24}$$

Step 4 : Substitute the equivalent fractions obtained in step 2 and 3.

$$\frac{5}{12} + \frac{1}{8} = \frac{10}{24} + \frac{3}{24}$$

Step 5 : The denominators of both fractions are same (Like denominators). Add the numerators.

$$\frac{10 + 3}{24} = \frac{13}{24}$$

Note : To add more than two fractions, repeat step 2 or step 3 for each of the fractions to convert them into equivalent fractions and then proceed with step 4 and 5.

$$\frac{5}{12} + \frac{1}{8} = \frac{13}{24}$$

Subtracting simple fractions, follow the below steps :

#1 : Fractions with like denominators can be subtracted by subtracting their numerators.

Example 6 : $\frac{3}{5} - \frac{1}{5} = \frac{3 - 1}{5} = \frac{2}{5}$

#2 : To subtract fractions with unlike denominators, convert the fractions to equivalent fractions with like denominators and follow #1. To convert them into equivalent fractions you can multiply the numerator and denominator with a common factor for each of the fractions to be subtracted separately and then subtract the fractions.

Method 1 :

Example 7 : $\frac{3}{5} - \frac{1}{2}$

Step 1 : $\frac{3}{5} = \frac{3 X 2}{5 X 2} = \frac{6}{10}$

Step 2 : $\frac{1}{2} = \frac{1 X 5}{2 X 5} = \frac{5}{10}$

Step 3 : $\frac{3}{5} - \frac{1}{2} = \frac{6}{10} - \frac{5}{10} = \frac{6 - 5}{10} \quad \frac{1}{10}$

$$\frac{3}{5} - \frac{1}{2} = \frac{1}{10}$$

Another method to subtract fractions with unlike denominators is by finding the **Least Common Multiple** (LCM) of the denominators and then follow the steps as described below in same order.

Method 2 :

Example 8 : $\frac{5}{12} - \frac{1}{8}$

Step 1 : Find the LCM of the denominators 12 , 8

$$\begin{array}{r|l} 2 & 12\ ,\ 8 \\ \hline 2 & 6\ ,\ 4 \\ \hline 3 & 3\ ,\ 2 \\ \hline 2 & 1\ ,\ 2 \\ \hline & 1\ ,\ 1 \end{array}$$

LCM = 2 X 2 X 3 X 2 = 24

Step 2 : Converting $\frac{5}{12}$ into an equivalent fraction with the denominator as 24 (LCM).

Divide the LCM value with the number in the denominator of the fraction to obtain the common factor.

$$\begin{array}{r} 2 \\ 12\overline{)\ 24} \\ -\ 24 \\ \hline 0 \end{array}$$

Common factor obtained is 2

$$\frac{5}{12} = \frac{5 X 2}{12 X2} = \frac{10}{24} \ ; \ \frac{5}{12} = \frac{10}{24}$$

Step 3 : Converting $\frac{1}{8}$ into an equivalent fraction with the denominator as 24 (LCM).

Divide the LCM value with the number in the denominator of the fraction to obtain the common factor.

$$\begin{array}{r} 3 \\ 8\overline{)\ 24} \\ -\ 24 \\ \hline 0 \end{array}$$

Common factor obtained is 3

$$\frac{1}{8} \quad \frac{1 X 3}{8 X 3} \quad \frac{3}{24} \quad \frac{1}{8} \quad \frac{3}{24}$$

Step 4 : Substitute the equivalent fractions obtained in step 2 and 3.

$$\frac{5}{12} - \frac{1}{8} = \frac{10}{24} - \frac{3}{24}$$

Step 5 : The denominators of both fractions are same (Like denominators). Subtract the numerators.

$$\frac{10 - 3}{24} = \frac{7}{24}$$

Note : To Subtract more than two fractions, repeat step 2 or step 3 for each of the fractions to convert them into equivalent fractions and then proceed with step 4 and 5.

$$\frac{5}{12} - \frac{1}{8} = \frac{7}{24}$$

Adding Mixed numbers :

#1 : Add the whole numbers together.

#2 : Add the fractional parts together. (Find a common denominator if necessary) (Follow the steps as described in the previous pages)

#3 : Write the whole number obtained from step 1 and the fraction obtained from step 2.

#4 : If the fractional part is an improper fraction, change it to a mixed number. Add the whole part of the mixed number to the original whole numbers. Rewrite the fraction in the lowest possible value.

Example 9 : $3\frac{2}{7} + 5\frac{1}{7}$

Step 1 : Add the whole part 3 from $3\frac{2}{7}$ to the whole part 5 from $5\frac{1}{7}$

$3 + 5 = 8$

Step 2 : Add the fractional part $\frac{2}{7}$ from $3\frac{2}{7}$ to the fractional part $\frac{1}{7}$ from $5\frac{1}{7}$

$$\frac{2}{7} + \frac{1}{7} = \frac{2 + 1}{7} = \frac{3}{7}$$

$$3\frac{2}{7} + 5\frac{1}{7} = 8\frac{3}{7}$$

Example 10 : $4\frac{1}{6} + 5\frac{5}{6}$

Step 1 : Add the whole part 4 from $4\frac{1}{6}$ to the whole part 5 from $5\frac{5}{6}$

$4 + 5 = 9$

Step 2 : Add the fractional part $\frac{1}{6}$ from $4\frac{1}{6}$ to the fractional part $\frac{5}{6}$ from $5\frac{5}{6}$

$\frac{1}{6} + \frac{5}{6} = \frac{1+5}{6} = \frac{6}{6}$ (Numerators are added for fractions from like denominators)

Step 3 : $4\frac{1}{6} + 5\frac{5}{6} = 9\frac{6}{6} = 10$

($\frac{6}{6}$ = one whole part, add this one whole part to 9 making it equal to 10)

Example 11 : $7\frac{1}{2} + 5\frac{3}{20}$

Step 1 : Add the whole part 7 from $7\frac{1}{2}$ to the whole part 5 from $5\frac{3}{20}$

$7 + 5 = 12$

Step 2 : Add the fractional part $\frac{1}{2}$ from $7\frac{1}{2}$ to the fractional part $\frac{3}{20}$ from $5\frac{3}{20}$

$\frac{1}{2} + \frac{3}{20}$ (The fractions has unlike denominators)

Finding the LCM of 2 and 20

2	2 ,20
2	1 ,10
5	1 , 5
	1 , 1

LCM of 2 and 20 = 2 X 2 X 5 = 20

Step 3 : Let's make $\frac{1}{2}$ as an equivalent fraction with a denominator of 20.

$$\begin{array}{r} 10 \\ 2\overline{)\,20} \\ -20 \\ \hline 0 \end{array}$$

Common factor is 10

$\frac{1}{2} = \frac{1 \text{ X } 10}{2 \text{ X } 10} = \frac{10}{20}$

Step 4 : The fractional part $\frac{3}{20}$ of $5\frac{3}{20}$ has the same common denominator as 20.

We do not need to convert $\frac{3}{20}$ into another equivalent fraction.

Remember : The fractions can vary from problem to problem and students need to follow step 3 for all the fractional parts to convert them to equivalent fractions.

Step 5 : $\frac{1}{2} + \frac{3}{20} = \frac{10}{20} + \frac{3}{20} = \frac{10 + 3}{20} = \frac{13}{20}$

Step 6 : $7\frac{1}{2} + 5\frac{3}{20} = 8\frac{13}{20}$

Subtracting Mixed numbers :

#1 : Subtract the whole numbers together.

#2 : Subtract the fractional parts together. (Find a common denominator if necessary) (Follow the steps as described in the previous pages)

#3 : Write the whole number obtained from step 1 and the fraction obtained from step 2.

#4 : If the fractional part is an improper fraction, change it to a mixed number. Add the whole part of the mixed number to the whole number obtained in step 1. Rewrite the fraction in the lowest possible value.

Example 12 : $9\frac{2}{7} - 5\frac{1}{7}$

Step 1 : Subtract the whole part 5 from $5\frac{1}{7}$ from the whole part 9 from $3\frac{2}{7}$

$9 - 5 = 4$

Step 2 : Subtract the fractional part $\frac{1}{7}$ from $5\frac{1}{7}$ from the fractional part $\frac{2}{7}$ from $3\frac{2}{7}$

$\frac{2}{7} - \frac{1}{7} = \frac{2 - 1}{7} = \frac{1}{7}$

Step 3 : $9\frac{2}{7} - 5\frac{1}{7} = 4\frac{1}{7}$

Example 13 : $7\frac{5}{6} - 5\frac{1}{6}$

Step 1 : Subtract the whole part 5 from $5\frac{1}{6}$ from the whole part 7 from $7\frac{1}{6}$

$7 - 5 = 2$

Step 2 : Subtract the fractional part $\frac{1}{6}$ from $5\frac{1}{6}$ to the fractional part $\frac{5}{6}$ from $7\frac{5}{6}$

$\frac{5}{6} - \frac{1}{6} = \frac{5-1}{6} = \frac{4}{6}$ (Numerators are added for fractions from like denominators)

Step 3 : $7\frac{5}{6} - 5\frac{1}{6} = 2\frac{4}{6} = 2\frac{2}{3}$

($\frac{4}{6} = \frac{2}{3}$ Equivalent fractions)

Example 14 : $7\frac{1}{2} - 5\frac{3}{20}$

Step 1 : Subtract the whole part 5 from $5\frac{3}{20}$ from the whole part 7 from $7\frac{1}{2}$

$7 - 5 = 2$

Step 2 : Subtract the fractional part $\frac{3}{20}$ from $5\frac{3}{20}$ to the fractional part $\frac{1}{2}$ from $7\frac{1}{2}$

$\frac{3}{20} - \frac{1}{2}$ (The fractions has unlike denominators)

Finding the LCM of 2 and 20

$$\begin{array}{r|l} 2 & 2\ ,20 \\ \hline 2 & 1\ ,10 \\ \hline 5 & 1\ ,\ 5 \\ \hline & 1\ ,\ 1 \end{array}$$

LCM of 2 and 20 = 2 X 2 X 5 = 20

Step 3 : Let's make $\frac{1}{2}$ as an equivalent fraction with a denominator of 20.

$$\begin{array}{r} 10 \\ 2\overline{)\,20} \\ -20 \\ \hline 0 \end{array}$$

Common factor is 10

$\frac{1}{2} = \frac{1 \times 10}{2 \times 10} = \frac{10}{20}$

Step 4 : The fractional part $\frac{3}{20}$ of $5\frac{3}{20}$ has the same common denominator as 20.

We do not need to convert $\frac{3}{20}$ into another equivalent fraction.

Remember : The fractions can vary from problem to problem and students need to follow step 3 for all the fractional parts to convert them to equivalent fractions.

Step 5 : $\frac{1}{2} - \frac{3}{20} = \frac{10}{20} - \frac{3}{20} = \frac{10 - 3}{20} = \frac{7}{20}$

Step 6 : $7\frac{1}{2} - 5\frac{3}{20} = 2\frac{7}{20}$

Example 15 : $5\frac{1}{7} - 3\frac{3}{7}$

Step 1 : Subtract the whole part 3 from $3\frac{3}{7}$ from the whole part 5 from $5\frac{1}{7}$

$5 - 3 = 2$

Step 2 : Subtract the fractional part $\frac{3}{7}$ from $3\frac{3}{7}$ from the fractional part $\frac{1}{7}$ from $5\frac{1}{7}$

We cannot subtract $\frac{3}{7}$ from $\frac{1}{7}$

So we need to rewrite the fraction $5\frac{1}{7}$

$5\frac{1}{7} = 4\frac{8}{7}$ (Remember in this fraction one whole part equals to seven. so when we take one whole part into the fraction form we need to add 7 to the value in the numerator which equals to 7 + 1 = 8)

Step 3 : Repeat step 1

$5\frac{1}{7} - 3\frac{3}{7} = 4\frac{8}{7} - 3\frac{3}{7}$

Subtract the whole part 3 from $3\frac{3}{7}$ from the whole part 4 from $4\frac{8}{7}$

$4 - 3 = 1$

Step 4 : Subtract the fractional part $\frac{3}{7}$ from $3\frac{3}{7}$ from the fractional part $\frac{8}{7}$ from $4\frac{8}{7}$

$$\frac{8}{7} - \frac{3}{7} = \frac{8-3}{7} = \frac{5}{7}$$

Step 5 : $5\frac{1}{7} - 3\frac{3}{7} = \frac{5}{7}$

Multiplying Fractions :

#1 : Verify if the fractions are in lowest possible values. If not convert them into lowest possible values.

#2 : Using cross simplification method simplify the fractions, meaning a numerator can be simplifies with a denominator only and vice versa.

#3 : Do not cross simplify numerator with a numerator value and denominator with a denominator value

#4 : Multiply the numerator with the remaining numerator values and the denominator with the denominator values

Remember : "Top times the top over the bottom times the bottom".
All the answers must be written in simplest form.

Example 16 : $\frac{6}{15} \text{ X } \frac{3}{10}$

$$\frac{\overset{2}{\cancel{6}}}{\underset{5}{\cancel{15}}} \text{ X } \frac{3}{10} = \frac{\overset{1}{\cancel{2}}}{5} \text{ X } \frac{3}{\underset{5}{\cancel{10}}} = \frac{1}{5} \text{ X } \frac{3}{5}$$

$$= \frac{1 \text{ X } 3}{5 \text{ X } 5} = \frac{3}{25}$$

Multiplying Mixed Numbers :

#1 : To multiply mixed numbers, convert them to improper fractions.

Converting mixed number to improper fractions :
Multiply the denominator of the fraction to the whole part and then add the product to the numerator.

Example 17 : $2\frac{1}{3} = \frac{2 \text{ X } 3 + 1}{3} = \frac{6 + 1}{3} = \frac{7}{3}$

#2 : Verify if the fractions are in lowest possible values. If not convert them into lowest possible values.

#3 : Using cross simplification method simplify the fractions, meaning a numerator can be simplifies with a denominator only and vice versa.

#4 : Do not cross simplify numerator with a numerator value and denominator with a denominator value

#5 : Multiply the numerator with the remaining numerator values and the denominator with the denominator values

Remember : "Top times the top over the bottom times the bottom".
All the answers must be written in simplest form.
All improper fractions must be change back to mixed numbers.

Note : MULTIPLICATION CAN BE WRITTEN WITH THE SYMBOLS X OR **.** IN BETWEEN, .

Dividing Fractions :

#1 : To divide fractions, convert the division problem into a multiplication problem. Do this by multiplying the first fraction by the reciprocal of the second fraction. In other words convert the division to multiplication and interchange the numerator and denominator of the second fraction.

Remember : "When two fractions we divide, flip the second and multiply."

Note : Don't forget to check for "cross simplification" when multiplying.
All answers must be written in simplest form.
All improper fractions must be change back to mixed numbers.

Example 18 : $\frac{6}{15} \div \frac{3}{10}$

$$\frac{6}{15} \div \frac{3}{10} = \frac{\overset{2}{\cancel{6}}}{\underset{5}{\cancel{15}}} \times \frac{10}{3} = \frac{2}{\underset{1}{\cancel{5}}} \times \frac{\overset{2}{\cancel{10}}}{3} = \frac{2}{1} \times \frac{2}{3}$$

$$= \frac{2 \times 2}{1 \times 3} = \frac{4}{3}$$

Dividing Mixed Numbers :

#1 : To divide mixed fractions, first change them to improper fractions.

#2 : To divide fractions, convert the division problem into a multiplication problem. Do this by multiplying the first fraction by the reciprocal of the second fraction. In other words convert the division to multiplication and interchange the numerator and denominator of the second fraction.

Remember : "When two fractions we divide, flip the second and multiply."

Note : Don't forget to check for "cross simplification" when multiplying.
All answers must be written in simplest form.
All improper fractions must be change back to mixed numbers.

Example 19 : $2\frac{6}{15} \div 5\frac{3}{10}$

$$2\frac{6}{15} \div 5\frac{3}{10} = \frac{2 \times 15 + 6}{15} \div \frac{5 \times 10 + 3}{10} = \frac{30 + 6}{15} \div \frac{50 + 3}{10}$$

$$= \frac{30 + 6}{15} \div \frac{50 + 3}{10}$$

$$= \frac{36}{15} \div \frac{53}{10}$$

$$= \frac{36}{15} \text{ X } \frac{10}{53}$$

(Remember when we change division to multiplication, flip the second fraction)

$$= \frac{\overset{12}{\cancel{36}}}{\underset{5}{\cancel{15}}} \text{ X } \frac{10}{53}$$

$$= \frac{12}{\underset{1}{\cancel{5}}} \text{ X } \frac{\overset{2}{\cancel{10}}}{53}$$

$$= \frac{12}{1} \text{ X } \frac{2}{53}$$

$$= \frac{12 \text{ X } 2}{1 \text{ X } 53}$$

$$= \frac{24}{53}$$

$$2\frac{6}{15} \div 5\frac{3}{10} = \frac{24}{53}$$

Decimals notes

The word standard means regular. Numbers in standard form are whole numbers or natural numbers.

Example : The number "six hundred twenty five" in standard form is 625.

To name a decimal from its standard form, follow these steps :

1. Name the number in front of the decimal. (Do not include the word "and").
2. The word "and" is used for the decimal point.
3. The number in the decimal part is similar to the number in front of the decimal.
4. Name the last place value given (of the digit farther to the right).

Rounding Decimals to a Given Place Value:

To round a decimal number to a given place value, look at the digit to the right of the desired place value and follow the rounding rules:

"5 and above, give it a shove! 4 or below, leave it alone!"

(The number in the desired place value gets "bumped up" to the next consecutive value if the digit to the right of it is 5 or more. The number in the desired place value does not change if the digit to the right of it has a value of 5 or less.)

The purpose of rounding is to provide *an* estimate and get an approximate value. Rounding involves losing some accuracy.

Example : If 5,953 people attend a Soccer game. We can approximately say 6,000 people watched the game.

Comparing and Ordering Decimals:

To order the given decimals, compare the digits of all decimals according to their place value.

Tip : Line the numbers up vertically according to their place value and compare from the left to the right. When comparing like place values from the left, the number with the higher digit is the larger number.

Also, added zeros to the right of a decimal number does not change the value of the decimal.

For example, 0.8 = 0.80 = 0.800 = 0.8000 = 0.80000

Naming/Reading Decimal :

Read the number to the left of the decimal. Say the word “and” for the decimal place. Read the number to the right of the decimal as you would read it if the number were on the left. End the name with the place value of the digit that is furthest to the right.

Converting Between Fractions and Decimals :

Fractions and decimals represent a certain “part of a whole”, all fractions can be written as decimals and terminating/repeating decimals can be written as fractions.

Multiplying Decimals :

To multiply decimals, first ignore the decimals and simply multiply the digits. Then count the total number of spaces from the right to the decimal (or digits to the right of the decimal) of both numbers; place the decimal that number of spaces from the right into your answer.

Dividing Decimals:

Divide the decimals as a whole numbers ignoring the decimals.

1. Count the number of digits of the dividend after the decimal.
2. Count the number of digits of the divisor.
3. Subtract the number obtained in step 1 from the number obtained instep 2.
4. Count the number of digits from the right to the number obtained in step 3.
5. Place the decimal point.

Integers notes

Integers are a group , or set of numbers that consist of “whole numbers and their opposites”

1. Natural numbers and whole numbers are subset of integers.
2. The set does not include fractions or decimals.
3. The set includes positive and negative numbers.
4. Integers include : $-\infty$..... , -5 , -4 , -3 , -2 , -1 , 0 , 1 , 2 , 3 , 4 , 5 ...... $+\infty$
5. Integers greater than zero are called positive integers.
6. Integers less than zero are called negative integers.
7. Zero is neither negative nor positive.
8. Negative integers are the numbers to the left of 0.
   Example : -5 , -4 , -3
9. Negative numbers have a negative (-) sigh in front of the number.
10. Positive integers are the numbers to the right of 0.
11. Positive numbers do not require the + sign in front.
12. If a number has no sign , it is a positive number.
    Example : 2 , 3 , 4 , 10 , 20
13. Negative numbers are frequently used in measurements.
    Example : To measure temperatures , depth etc
    4^0 C below zero degree celsius is represented as -4^0
    100 ft below sea level is represented as -100 ft.
14. Arrows on a number line represent the numbers continuing for ever.
15. Positive numbers are represented on the right side of zero on the number line.
16. Negative numbers are represented on the left side of zero on the number line.
17. Number are placed at equal intervals on the number line. Not necessarily one unit.

Integers can be represented on a number line as below

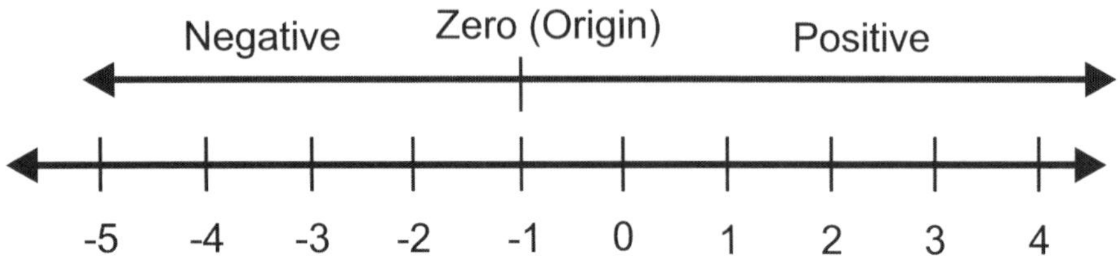

Absolute value :

The number line can be used to find the absolute value. The absolute value of an integer is the distance the number is from zero on the number line.

The absolute value of 2 is 2. Using the number line , 2 is a distance of 2 to the right of zero.The absolute value of -2 is also 2. Again using the number line , the distance from -2 to zero is 2. A measure of distance is always positive.

The symbol for absolute value of any number , x , is | x |.

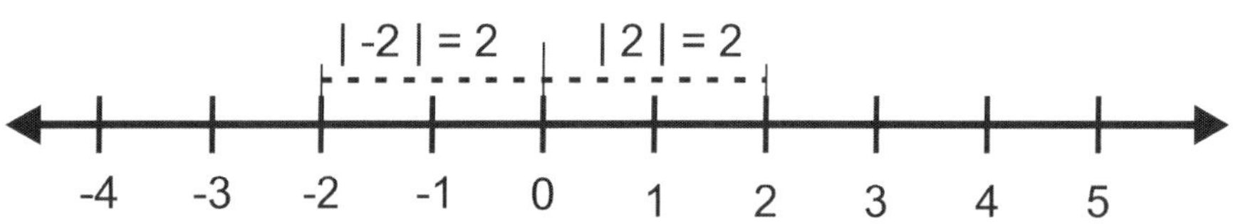

Opposite integers :

The opposite of an integer is the number that is at the same distance from zero in the opposite direction. Every integer has an opposite value, but the opposite of zero is itself.

The opposite of -4 is 4 because it is located the same distance from zero as 4 is , but in opposite direction.

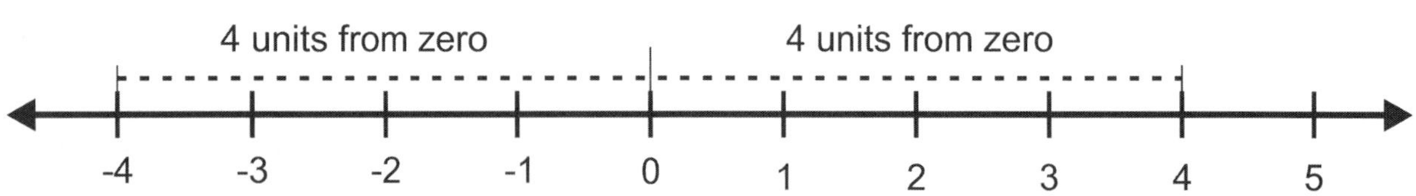

Adding integers using a number line :

The number line is visual representation to understand the addition of positive and negative numbers. Start with the one value on the number line, then add the second value. If the second value (that is added) is positive, we move to the right that many spaces.

If the second value (that is added) is negative, we move to the left that many spaces. The value where we land on the number line is the solution for the addition of two integers.

Example 1 : (-4) + (5) = 1
Start at the first number, -4, and travel 5 units to the right.

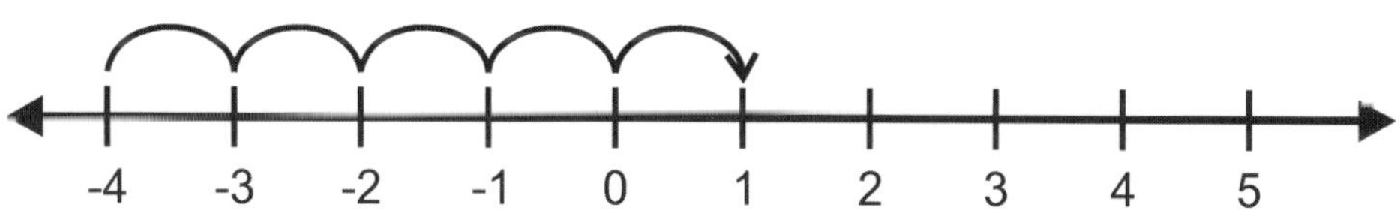

Example 2 : (5) + (-7) = -2
Start at the first number, 5, and travel 7 units to the left.

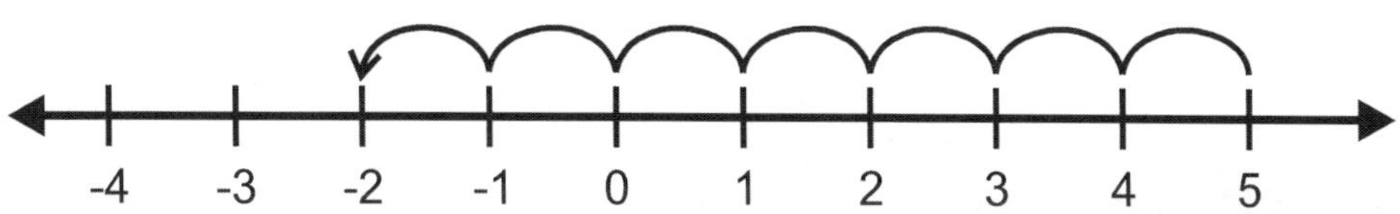

Adding integers using the rules :

Rules for adding integers :

If the signs are the same, add their absolute values, and keep the common sign.

If the signs are different, find the difference between the absolute values of the two numbers, and keep the sign of the number with the greater numerical value.

To the Tune of "Row Your Boat"

Same signs add and keep
Different signs subtract
Keep the sign of the greater digits
then you'll be exact

Subtacting integers using a number line :

A number line is helpful in understanding subtraction of positive and negative values. Start with the first value on the number line, then subtract the second value. If the second value (that is subtracted) is positive, we move to the left that many spaces.

If the second value (that is subtracted) is negative, we move to the right that many spaces. This is because subtraction a negative is the same as adding.
The value where we end on the number line is the answer.

Example 1 : (-2) + (5) = 3
Start at the first number, -2, and travel 5 units to the right.

Subtacting integers using the rules :

Every subtraction problem can be written as an additional problem. When we subtract two integers, just ADD THE OPPOSITE. Subtracting a positive is the same as adding a negative. Subtracting a negative is the same as adding a positive.

Multiplying Integers :

Multiplying integers is same as multiplying whole numbers, except we must keep track of the signs associated to the numbers.

To multiply signed integers, always multiply the absolute values and use these rules to determine the sign of the product value

When we multiply two integers with the same signs, the result is always a positive value.

Positive number X Positive number = Positive number

Negative number X Negative number = Positive number

When we multiply two integers with different signs, the result is always a negative value.

Positive number X Negative number = Negative number

Negative number X Positive number = Negative number

Positive X Positive : 7 X 6 = 42 — negative X negative : -7 X -6 = 42

Positive X negative : 7 X -6 = -42 — negative X Positive : -7 X 6 = -42

Dividing Integers :

Division of integers is similar to the division of whole numbers, except we must keep track of the signs associated.

To divide signed integers, we must always divide the absolute values and use the below rules to find the quotient value.

When we divide two integers with the same signs, the result is always a positive value.

Positive ÷ Positive = Positive

Negative ÷ Negative = Positive

When we divide two integers with opposite signs, the result is always a negative value.

Positive ÷ Negative = Negative

Negative ÷ Positive = Negative

Examples :

Positive ÷ Positive : 81 ÷ 9 = 9

negative ÷ negative : -81 ÷ -9 = 9

Positive negative : 81 ÷ -9 = -9

negative Positive : -81 ÷ 9 = -9

Golden Rules of Integers :

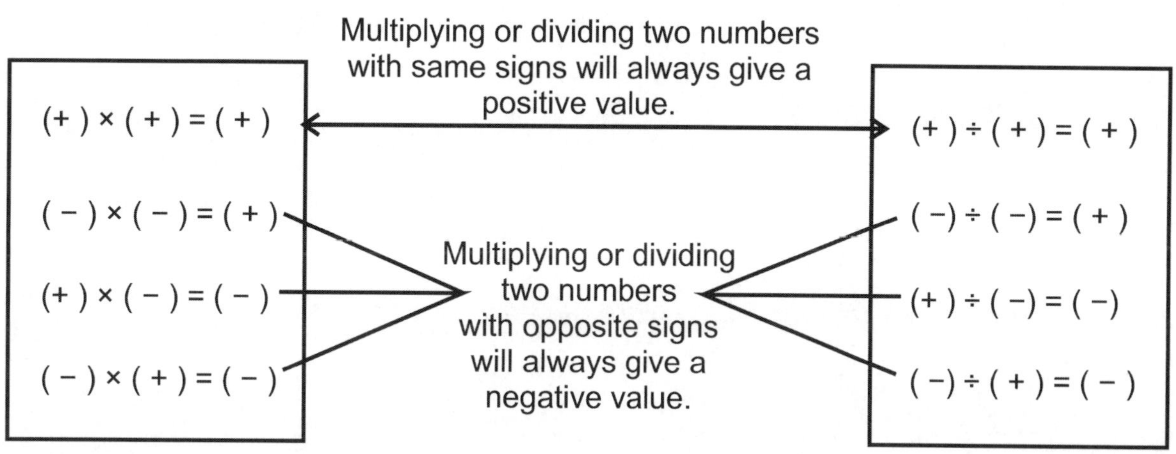

Introduction

Basic rules of number system involves addition, subtraction , multiplication and division. As the topic of "Numbers" involves some more useful concepts like LCM and GCD, we shall study them in this chapter. For finding LCM and GCD, the divisibility rules are useful in one way or the other. Hence let us start the learning the divisibility rules.

Divisibility:

In general, if two natural numbers a and b are such that, when 'a' is divided by 'b', a remainder of zero is obtained, we say that 'a' is divisible by 'b'.

For example, 12 is divisible by 3 because 12 when divided by 3, the remainder is zero.

Also, we say that 12 is not divisible by 5, because 12 when divided by 5, it leaves a remainder 2.

Tests of Divisibility:

We now study the methods to test the divisibility of natural numbers with 2, 3, 4, 5, 6, 8, 9 and 11 without performing actual division.

Test of Divisibility by 2:

A natural number is divisible by 2, if its units digit is divisible by 2, i.e., the units place is either 0 or 2 or 4 or 6 or 8.

Examples : The numbers 4096, 23548 and 34052 are divisible by '2' as they end with 6, 8 and 2 respectively.

Test of Divisibility by 3:

A natural is divisible by 3 if the sum of its digits is divisible by 3.

Example: Consider the number 2143251. The sum of the digits of 2143251 (2 + 1 + 4 + 3 + 2 + 5 + 1) is 18.

As 18 is divisible by 3, the number 2143251, is divisible by 3.

Test of Divisibility by 4:

A natural number is divisible by 4, if the number formed by its last two digits is divisible by 4.

Examples : 4096, 53216, 548 and 4000 are all divisible by 4 as the numbers formed by taking the last two digits in each case is divisible by 4.

Test of Divisibility by 5:

A natural number is divisible by 5, if its units digit is either 0 or 5.

Examples : The numbers 4095 and 235060 are divisible by 5 as they have in their units place 5 and 0 respectively.

Test of Divisibility by 6:

A number is divisible by 6, if it is divisible by both 2 and 3.

Examples : Consider the number 753618

Since its units digit is 8, so it is divisible by 2. Also its sum of digits = 7 + 5 + 3 + 6 + 1 + 8 = 30, As 30 is divisible by 3, so 753618 is divisible by 3.
Hence 753618 is divisible by 6.

Test of Divisibility by 8:

A number is divisible by 8, if the number formed by its last three digits is divisible by 8.

Examples : 15840, 5432 and 7096 are all divisible by 8 as the numbers formed by last three digits in each case is divisible by 8.

Test of Divisibility by 9:

A natural number is divisible by 9, if the sum of its digits is divisible by 9.

Examples :

(i) Consider the number 125847.
Sum of digits = 1 + 2 + 5 + 8 + 4 + 7 = 27. As 27 is divisible by 9, the number 125847 is divisible by 9.

(ii) Consider the number 145862.
Sum of digits = 1 + 4 + 5 + 8 + 6 + 2 = 26. As 26 not divisible by 9, the number 145862 is not divisible by 9.

Test of Divisibility by 11:

A number is divisible by 11, if the difference between the sum of the digits in odd places and sum of the digits in even places of the number is either 0 or a multiple of 11.

Examples :

(i) Consider the number 9582540
Now (sum of digits in odd places) - (sum of digits in even places)
$=(9+8+5+0)-(5+2+4)$
= 11, which is divisible by 11.
Hence 958254 is divisible by 11.

(ii) Consider the number 1453625
Now, (sum of digits at odd places) - (sum of digits at even places)
$-(1+5+6 \mid 5)-(4 \mid 3 \mid 2)$
$=17-9=8$, which is not divisible by 11

Some Additional Results :

If a natural number N is divisible by two natural numbers a and b, then N is divisible by the product of a and b, if and only if a and b are co-primes.

Examples :

(i) 345 is divisible by 3 as well as by 5, as 3 and 5 are co-primes, 345 is divisible by 15.
(ii) 120 is divisible by 8 and 10, but it is not divisible by 80.

Factors and Multiples :

Lets learn the concepts of factors and multiples.

Definition :

If 'b' divides 'a' leaving a zero remainder, then 'b' is called a factor or divisor of 'a' and 'a' is called the multiple of 'b'.
For example, $6=2\ \ 3$
Here 2 and 3 are factors of 6 (or) 2 and 3 are divisors of 6.
And, 6 is a multiple of 3, 6 is a multiple of 2.

Examples : (i) The factors of $24=\{1, 2, 3, 4, 6, 8, 12, 24\}$
(ii) The factors of $256=\{1, 2, 4, 8, 16, 32, 64, 128, 256\}$

Observations :

One is the factor of every natural number and it is the least of the factors of any natural number.
Every natural number is the factor of itself and it is the greatest.

Some Additional Results :

Unique Prime Factorisation Theorem : "Any natural number greater than 1 can be divided into a prime number or a composite number."

For example:

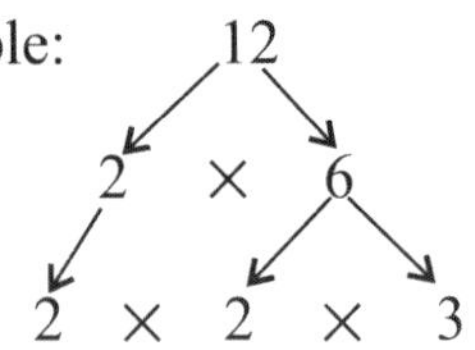

6 is a composite number

All factors are prime numbers.

If a is a composite number of the form $a = b^p c^q d^r$ ….. where b, c, d … are district prime factors, then the number of factors of $a = (p + 1)(q + 1)(r + 1)$ ….

If a is a composite number of the form $a = b^p c^q d^r$ …. where b, c, d …. are district prime factors, then the sum

of all the factors of a $= \frac{(b^{p-1} - 1)}{(b - 1)} \times \frac{(c^{q-1} - 1)}{(c - 1)} \times \frac{(d^{r-1} - 1)}{(d - 1)}$ ……

Perfect Numbers :

A number for which sum of all its factors is twice the number itself is called a perfect number.

Observation :

Euler proved that if $2^k - 1$ is a prime number, then $2^{k-1}(2^k - 1)$ is a perfect number.
A perfect number can never be a prime number.

Examples :

(i) Consider the composite number 6.
Factors of 6 = {1, 2, 3, 6}
Sum of factors = (1 + 2 + 3 + 6) = 12
Clearly, the sum of the factors of 6 is twice the number itself.
Hence 6 is a perfect number.

Example :

(ii) Consider the composite number 48
Factors of 48 = {1, 2, 3, 4, 6, 8, 12, 16, 24, 48}
Sum of factors = 1 + 2 + 3 + 4 + 6 + 8 + 12 + 16 + 24 + 48
= 124 2 48
Clearly, 48 is not a perfect number.

Greatest Common Divisor [GCD] (or) Greatest Common Factor [GCF] (or) Highest Common Factor [HCF]

Definition :

"The greatest common factor of two or more natural numbers is the largest factor in the set of common factors of those numbers." In other words, the GCD (or) GCF of two or more numbers is the largest number that divides each of them exactly.

Example : Find the GCF of 72 and 60.

Solution : Let the set of factors of 72 be A.
A = {1, 2, 3, 4, 6, 8, 9, 12, 18, 24, 36, 72}
Let the set of factors of 60 be B.
B = {1, 2, 3, 4, 5, 6, 10, 12, 15, 20, 30, 60}
The set of common factors for 72 and 60 is A B = {1, 2, 3, 4, 6, 12}
The greatest element in this set is 12
The GCF (or) GCD for 72 and 60 is 2.

Observations :
If two numbers have no factors in common, then their GCF is unity.
i.e., GCF of prime numbers and co-prime numbers is unity.

Methods of finding GCF :

Factors Method :

When the numbers whose GCF has to be found are relatively small, this is the best suited method. Here we resolve the given numbers into their prime factors and find out the largest factor in the set of common factors to given numbers.

This method can be easily applied to any number of numbers.

Examples : (i) Find the GCF of 24 and 36.

Solution :

Resolving given numbers into product of prime factors, we have
36 = ② × ② × ③ × 3
24 = 2 × ② × ② × ③
The common factors to both the numbers are circled.
Now GCF = product common factors of given numbers = 2 × 2 × 3 = 12
GCF (24, 36) = 12

(ii) Find the GCF of 12, 18 and 24.

Solution :

Resolving given numbers into product of prime factors;

12 = ② × 2 × ③

18 = ② × 3 × ③

24 = ② × 2 × 2 × ③

GCF = product of common factors of 12, 18 and 24

= 2 × 3 = 6

GCF = 6

Division Method :

When the numbers whose GCF has to be found are very large, it is time consuming to write down sets of common factors to given numbers;

In this case, we use the method of Long Division. This method was proposed by Euclid and the following steps are involved in it.

Step 1 : Divide the larger number by the smaller number. If the remainder is zero, the divisor is the GCF, otherwise not.

Step 2 : Let the divisor in step 2 be the dividend now, and the remainder of step 1 become the divisor of step 2. Again, if the remainder is zero, the divisor is GCD. Otherwise, step 2 has to be repeated.

Example : Find the GCF of 64 and 56

Solution :

64 divided by 56, quotient is 1 and remainder is 8. Because the reminder is not zero; 56 is not the GCD.

Proceeding further, as mentioned in step 2, 56 is dividend and 8 is divisor. The quotient is 3 and remainder is zero.

Because the remainder is zero, the divisor 8 is the GCD.

GCF of three numbers using division method:

The GCF of 3 numbers is found out by finding GCF of any 2 numbers and GCF of the remaining number with the GCF obtained above.

i.e., GCF (a, b, c) = GCF [GCF (a, b), c].

This process can be extended to any number of numbers.

Example : Find the GCF of 25, 45 and 75.

Solution : Let us first find the GCF of 25 and 45

```
25) 45 (1
    25
    20) 25 (1
        20
         5) 20 (4
            20
             0
```

0

GCF (25, 45) = 5
The GCF of (5, 75) is 5.
The GCF of 25, 45 and 75 is 5.

Observations :

The process of dividing the divisor with quotient is to be repeated until remainder as zero is obtained

If the zero remainder is obtained when the divisor is 1, then the GCD is '1'.

GCD is '1', means that two numbers are relatively prime or co-prime; because they do not have any factor in common other than 1. For eg; 12 and 13 are co-primes.

Some Additional Results :

The largest number which divides p, q and r to give remainders of s, t and u respectively will be the GCD of the three numbers $(p-s)$, $(q-t)$ and $(r-u)$.

The largest number which divides the numbers p, q and r and gives the same remainder in each case will be the GCD of the differences of two or the three numbers $(p-q, q-r, p-r)$.

Least Common Multiple [LCM] :

Definition :

"The least common multiple of two or more natural numbers is the least of their common multiples". In other words, the LCM of two or more numbers is the least number which can be divided exactly by each of the given numbers.

Note : If the set of common multiples is denoted by C, then N and the number of elements in C is infinite and the least element in C is their LCM.

Example : Find L.C.M. of 24 and 36.

Solution : Resolving 24 and 36 into product of prime factors
24 = ② × ② × 2 × ③
36 = ② × ② × ③ × 3

The common prime factors of 24 and 36 are 2, 2 and 3. (which are circled)

The remaining prime factors of 24 is 2. (which is not circled).

The remaining prime factors of 36 is 3. (which is not circled).

LCM = Common factors of 24 the prime factors left in 24 the prime factors left in 36
= 2 × 2 × 3 × 2 × 3
= 72

Methods of finding LCM :

Factors Method :

Here the given numbers are decomposed into product of prime factors; from which, the least common multiple is found by multiplying the terms containing factors of numbers raised to their highest powers.

Example : Find the LCM of 32 and 24.

Solution : Resolving given numbers into product of common factors, we have
$32 = 2^5$; $24 = 2^3 \times 3$
LCM = product of terms containing highest powers of factors 2, 3
$= 2^5 \; 3 = 96$

LCM of three numbers using factors method:

The method above can be extended in a similar way to three numbers. This is illustrated below:

Example : Find the LCM of 12, 48 and 36

Solution : Resolving given numbers into product of common factors, we have
$12 = 2^2 \times 3^1$; $48 = 2^4 \times 3^1$; $36 = 2^2 \times 3^2$. Then their LCM $= 2^2 \times 3^2 = 16 \times 9 = 144$

Synthetic Division Method of Finding LCM:

LCM of numbers can also be found using synthetic division method. This is illustrated below:

Example : Find the LCM of 144 and 156.

Solution : Using synthetic division, we have:

2	144, 156
2	72, 78
3	36, 39
	12, 13

LCM = 2 2 3 12 13
= 1872

Examples : Find the LCM of 12, 18 and 24

Solution : Using synthetic division;

2	12, 18, 24
2	6, 9, 12
2	3, 9, 6
	1, 3, 2

LCM = 2 2 2 1 3 2 = 48

Some Additional Results :

Any natural N number which when divided by p, q or r leaving the same remainder s in each case will be of the form N = K (LCM of p, q, r) + s, where K = 0, 1, 2, 3 ....

Any natural number N which when divided by p, q and r leaves respective remainders of s, t and u where (p – s) = (q – t) = (r – u) = v (say), then it will be of the form
N = k (LCM of p, q and r) – v, where k = 1, 2, 3 ....

A natural N number which when divided by p and q leaves remainders r and s, will be of the form N = k (LCM of p and q) + n when n is the smallest integer solution for the equations $n = pm_1 + r$ and $n = qm_2 + s$, where m_1 and m_2 are natural numbers.

Relationship between LCM and GCF :

The LCM and GCF of two given numbers are related to the given numbers by the following relationship.
Product of the numbers = LCM × GCF

where, LCM denotes the LCM of the given numbers and GCF denotes the GCF of the given numbers.

Example : Consider two numbers, 24 and 36.
These can be resolved into product of prime factors as below :

$24 = 2^3 \times 3$

$36 = 2^2 \times 3^2$

Now LCM (24, 36) $= 2^3 \times 3^2 = 72$

GCF (24, 36) $= 2^2 \times 3 = 12$

Now; Product of numbers $= 24 \times 36 = 2^5 \times 3^3 = 864$
Product of LCM and GCF $= 72 \times 12 = 2^5 \times 3^3 = 864$

Clearly, Product of the numbers = Product of the LCM and GCF.

Relatively Prime Numbers :

Definition :

If two numbers do not have any common factors other than 1, then they are called relatively prime numbers or co-prime numbers.

Concept : We know, every number has at least two factors, 1 and itself. If it has more than two factors, it is a composite number and if it does not have any factor except 1 and itself, it is a prime number.

But if two numbers (prime or composite) are such that they have only one common factor '1' are called relatively prime.

Example : Consider three numbers 8, 18 and 25
Now,
A, The set of factors of 8 = {1, 2, 4, 8}
B, The set of factors of 18 = {1, 2, 3, 6, 9, 18}
C, The set of factors of 25 = { 1, 5, 25}

Now, A B = {1, 2} ; B C = {1} ; C A = {1}

Clearly the common factors for both (18, 25) and (8, 25) is 1 only.

They are generally written as (18, 25) = 1 and (8, 25) = 1

Note : Here; 18, 25 and 8 are not prime numbers (composite) but they are relatively prime numbers.

Observations :

The G.C.D. of two relatively prime numbers is 1 and their LCM is product of the numbers

Any two prime numbers are always relatively prime to each other.

Two relatively prime numbers need not be prime numbers.

LCM and GCF of Fractions :

The LCM and GCF of fractions can be determined by the following relations :

$$\text{LCM of fractions} = \frac{\text{LCM of numerators}}{\text{GCD of denominators}}$$

$$\text{GCF of fractions} = \frac{\text{GCD of numerators}}{\text{LCM of denominators}}$$

Examples : Find the GCD and LCM of $\frac{4}{5}$, $\frac{2}{5}$ and $\frac{3}{4}$

Solutions : $\text{LCM}\left(\frac{4}{5}, \frac{2}{5}, \frac{3}{4}\right) = \frac{\text{LCM }(4, 2, 3)}{\text{HCF }(5, 5, 4)} = \frac{4\times3}{5}$

$$\text{GCD }\left(\frac{4}{5}, \frac{2}{5}, \frac{3}{4}\right) = \frac{\text{HCF }(4, 2, 3)}{\text{LCM }(5, 5, 4)} = \frac{1}{5\times4} = \frac{1}{20}$$

Introduction of Exponents

Whenever any integer, let us say x, is added n times, the result obtained will be equal to n times x i.e. nx. But in case, if the integer x is multiplied for n times, the result obtained will be equal to x^n (which is called as exponential form). The problems relating to these will be studied under 'Exponents'. We shall look at the rules/properties pertaining to these exponential numbers in this chapter.

Rational Exponents and Radicals

If 'a' is any real number and 'n' is a positive integer, then the product a × a × a × ---- n times is represented by the notation a^n. This notation is referred to as exponential form. In the above notation, a is called the base and n is called the power or exponent or index (plural of index is indices). a^n is read as 'n^{th} power of a' or 'a to the power n'.

Example: 6 × 6 × 6 × 6 × 6 × 6 × 6 can be written as 6^7. Here 6 is called as base and 7 is called as index (or exponent).

For a non-zero rational number 'a' with a negative integral exponent 'm' the following result can be observed.

$a^m = a^{-n} = a^{-1} \times a^{-1} \times a^{-1} \times a^{-1} \times ---- \times ---$ n times

$= \frac{1}{a} \times \frac{1}{a} \times \frac{1}{a} \times \frac{1}{a}$ ---- n times $= \left(\frac{1}{a}\right)^n = \frac{1}{a^n}$

Example: $6^{3} = \left(\frac{1}{6}\right)^3$

Rational Indices:

1. n^{th} root of a :

A real number x is said to be the nth root of a if $x^n = a$; where a is any real number and n is a positive integer.

<u>Parts of the exponent:</u>

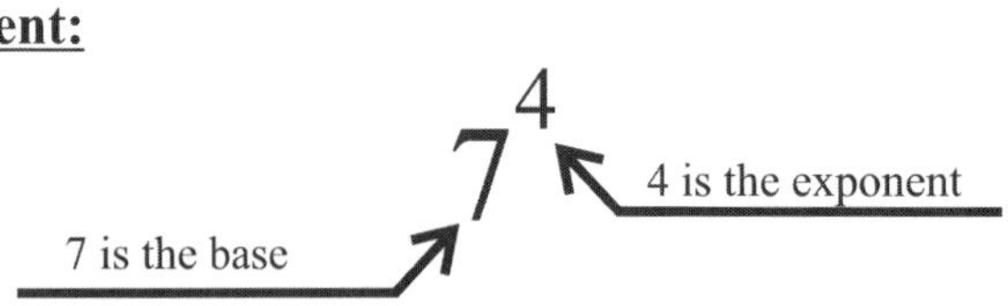

This is read as "Seven to the fourth power"

The nth root of a can be represented as $a^{1/n}$ or $\sqrt[n]{a}$. Here $a^{1/n}$ is called exponential form and the form $\sqrt[n]{a}$ is called radical form. The sign $\sqrt[n]{\ }$ is called radical sign and $\sqrt[n]{a}$ is called radical. The number n is a positive integer is called the index of radical and a is called the radicand.

Example: We know that $32 = 2^5$. So we can say that 2 is 5th root of 32. It is written as $32^{1/5} = 2$ or $\sqrt[5]{32} = 2$
Similarly $\sqrt[3]{64} = 3$, $\sqrt[4]{625} = 5$, $\sqrt[6]{64} = 2$, etc.

Note:

(i) If n is negative as in case $64^{-\frac{1}{3}}$, we write the radical form as follows;

$$64^{-\frac{1}{3}} = \left(\frac{1}{64}\right)^{\frac{1}{3}} = \sqrt[3]{\frac{1}{64}}.$$

The radical form of $64^{-\frac{1}{3}}$ should not be taken as $\sqrt[3]{64}$ as in the radical form $\sqrt[n]{a}$ of n is a positive integer.

i.e., $a^{1/n} = \sqrt[n]{\frac{1}{a}}$, where n is a positive integer.

(ii) $\sqrt[n]{a}$ is positive for a > 0 and n being a positive integer.
Example: $\sqrt[5]{32} = 2$, $\sqrt[6]{64} = 2$, $\sqrt[3]{27} = 3$, etc.

(iii) $\sqrt[n]{a}$ is negative for a < 0 and n being any odd positive integer.
Example: $\sqrt[3]{8} = -2$, $\sqrt[7]{128} = -2$, $\sqrt[5]{243} = -3$, etc.

(iv) $\sqrt[n]{a}$ does not exist in set of real numbers, for a < 0 and n being even positive integer.
Example: $\sqrt[2]{16}$, $\sqrt[4]{256}$, $\sqrt[6]{64}$ etc doesn't exist.

Each positive number has two square roots, one positive and the other negative.

Example: $\sqrt{36} = 6$ or -6 (since $6^2 = (-6)^2 = 36$).

If a is a positive rational number and n = p/q is a positive rational exponent, then we can define $a^{p/q}$ in two ways.

(1) $a^{\frac{p}{q}}$ is the qth root of a^p, i.e. $a^{\frac{p}{q}} = \left(a^p\right)^{\frac{1}{q}}$

(2) $a^{\frac{p}{q}}$ is the pth power of qth root of a, i.e. $a^{\frac{p}{q}} = \left(a^{\frac{1}{q}}\right)^p$.

Laws of indices:

For all real numbers a and b and all rational numbers m and n, we have

(i) $a^m \times a^n = a^{m+n}$

Examples: (1) $2^3 \times 2^6 = 2^{3+6} = 2^9$

(2) $\left(\frac{5}{6}\right)^4 \times \left(\frac{5}{6}\right)^5 = \left(\frac{5}{6}\right)^{4+5} = \left(\frac{5}{6}\right)^9$

(3) $5^{2/3} \times 5^{4/3} = 5^{(2/3+4/3)} = 5^{6/3} = 5^2$

(4) $2^3 \times 2^4 \times 2^5 \times 2^8 = 2^{(3+4+5+8)} = 2^{20}$.

(5) $\left(\sqrt{7}\right)^3 \times \left(\sqrt{7}\right)^{\frac{5}{2}} = \left(\sqrt{7}\right)^{3+\frac{5}{2}} = \left(\sqrt{7}\right)^{\frac{11}{2}}$

(ii) $a^m \div a^n = a^{m-n}, a \neq 0$

Examples: (a) $7^8 \div 7^3 = 7^{8-3} = 7^5$

(b) $\left(\frac{7}{3}\right)^9 \div \left(\frac{7}{3}\right)^5 = \left(\frac{7}{3}\right)^{9\ \ 5} = \left(\frac{7}{3}\right)^4$

(c) $9^{\frac{2}{3}} \div 9^{\frac{1}{6}} = 9^{\left(\frac{2}{3}-\frac{1}{6}\right)} = 9^{\left(\frac{4-1}{6}\right)} = 9^{\frac{3}{6}} = 9^{\frac{1}{2}}$

(d) $\left(\frac{5}{7}\right)^{\frac{8}{9}} \div \left(\frac{5}{7}\right)^{\frac{1}{3}} = \left(\frac{5}{7}\right)^{\left(\frac{8}{9}-\frac{1}{3}\right)} = \left(\frac{5}{7}\right)^{\frac{8-3}{9}} = \left(\frac{5}{7}\right)^{\frac{5}{9}}$

Note: $a^n \div a^n = 1$
or $a^{n-n} = a^0 = 1$
$\therefore a^0 = 1, a \neq 0$

(iii) $(a^m)^n = a^{m \times n}$

Examples: (a) $(5^2)^3 = 5^{2 \times 3} = 5^6$

(b) $\left[\left(\frac{2}{3}\right)^4\right]^5 = \left(\frac{2}{3}\right)^{4\times5} = \left(\frac{2}{3}\right)^{20}$

(c) $\left[\left(\frac{5}{7}\right)^{\frac{2}{3}}\right]^{\frac{9}{8}} = \left(\frac{5}{7}\right)^{\left(\frac{2}{3}\times\frac{9}{8}\right)} = \left(\frac{5}{7}\right)^{\frac{3}{4}}$

(iv) $\left(\frac{a}{b}\right)^n = \frac{a^n}{b^n}$

Example: $\left(\frac{4}{5}\right)^7 = \frac{4^7}{5^7}$

Note: Conversely we can write $\left(\frac{a^n}{b^n}\right) = \left(\frac{a}{b}\right)^n$

Example: $\frac{8}{27} = \frac{2^3}{3^3} = \left(\frac{2}{3}\right)^3$

(v) $(ab)^n = a^n \times b^n$

Examples: (a) $20)^5 = (4 \times 5)^5 = 4^5 \times 5^5$
(b) $(42)^7 = (2 \times 3 \times 7)^7 = 2^7 \times 3^7 \times 7^7$

Note: Conversely we can write $a^n \times b^n = (ab)^n$

Examples: (a) $4^8 \times 5^8 = (4 \times 5)^8 = 20^8$

(b) $\left(\frac{2}{3}\right)^5 \times \left(\frac{9}{8}\right)^5 = \left(\frac{2}{3} \times \frac{9}{8}\right)^5 = \left(\frac{3}{4}\right)^5$

(vi) $a^{-n} = \frac{1}{a^n}$, $a \neq 0$

Example: $2^{-4} = \frac{1}{2^4}$, $5^{-1} = \frac{1}{5}$

Note: $a^{-1} = \frac{1}{a^1} = \frac{1}{a}$

(vii) $\left(\frac{a}{b}\right)^{-n} = \left(\frac{b}{a}\right)^n$

Examples: (a) $\left(\frac{5}{9}\right)^3 = \left(\frac{9}{5}\right)^{-3}$

(b) $\left(\frac{1}{5}\right)^{-1} = \left(\frac{5}{1}\right)^1 = 5$

Note: $\left(\frac{1}{a}\right)^{-1} = \left(\frac{a}{1}\right)^1 = a$

(viii) If $a^m = a^n$, then $m = n$, where $a \neq 0$, $a \neq 1$

Examples: (a) If $5^p = 5^3 \Rightarrow p = 3$

(b) If $4^p = 256$

$4^p = 4^4 \Rightarrow p = 4$

(ix) For positive numbers a and b, if $a^n = b^n$, $n \neq 0$, then $a = b$ (when n is odd)

Examples: (a) If $5^7 = p^7$, then clearly $p = 5$.

(b) If $(5)^{2n-1} = (3 \times p)^{2n-1}$, then clearly $5 = 3p$ or $p = 5/3$

(x) If $p^m \times q^n \times r^s = p^a q^b r^c$, then m =a, n = b, s = c, where p, q, r are different primes.

Examples: (a) If $40500 = 2^a \times 5^b \times 3^c$, then find $a^a \times b^b \times c^c$

$$\begin{array}{r|l} 2 & 40{,}500 \\ \hline 2 & 20{,}250 \\ \hline 5 & 10{,}125 \\ \hline 5 & 2{,}025 \\ \hline 5 & 405 \\ \hline 3 & 81 \\ \hline 3 & 27 \\ \hline 3 & 9 \\ \hline & 3 \end{array}$$

$\therefore 40500 = 2^2 \times 5^3 \times 3^4 = 2^a \times 5^b \times 3^c$

$\therefore a = 2, b = 3, c = 4$, [Using the above law].

$\therefore a^a \times b^b \times c^c = 2^2 \times 3^3 \times 4^4 = 27{,}648$

Example 8 : (a) $20)^5 = (4 \times 5)^5 = 4^5 \times 5^5$

(b) $(42)^7 = (2 \times 3 \times 7)^7 = 2^7 \times 3^7 \times 7^7$

Note: Conversely we can write $a^n \times b^n = (ab)^n$

Example 9 : (a) $((5)^3)^2 = 5^{3X2} = (5)^6 = 5\times5\times5\times5\times5\times5 = 15625$

(b) $(2)^3 \times (2)^5 = (2)^{3+5} = (2)^8 = 2\times2\times2\times2\times2\times2\times2\times2 = 256$

(c) $(7)^0 = 1$ Any base value rise to the power zero is always equal to 1

(d) $(3)^{-4} = \dfrac{1}{(3)^4} = \dfrac{1}{3\times3\times3\times3} = \dfrac{1}{81}$

(d) $\dfrac{(8)^7}{(8)^5} = (8)^{7-5} = (8)^2 = 64$

(e) $\dfrac{(9)^4}{(9)^7} = (9)^{4-7} = (9)^{-3} = \dfrac{1}{(9)^3} = \dfrac{1}{9\times9\times9} = \dfrac{1}{729}$

(f) $(2)^4 = 2\times2\times2\times2 = 16$

(g) $(-2)^4 = -2\times-2\times-2\times-2 = 16$

(h) $-(2)^4 = -(2\times2\times2\times2) = -16$

(i) $-(2)^3 = -(2\times2\times2) = -8$

(j) $(-2)^3 = -2\times-2\times-2 = -8$

(k) $-(-2)^3 = -(-2\times-2\times-2) = -(-8) = 8$

Tip 1 : When the exponent is an even number the simplified value is always positive, when the base has a positive or negative value.

Tip 2 : When the exponent is an odd number the simplified value is always positive, when the base has a positive value.

Tip 3 : When the exponent is an odd number the simplified value is always negative, when the base has a negative value.

The percentage symbol is a representation of percentage. In statistics, percentages are often left in their base form of 0 - 1, where 1 represents the whole. We multiply the decimal by a factor of 100 to find the percentage.

Percent Equation form :

Percentages can be setup as proportions. The parts of a percent can be found by setting up a proportion as below.

$$\frac{\text{is}}{\text{of}} = \frac{\text{percent}}{100} \quad \text{OR} \quad \frac{\text{part}}{\text{total}} = \frac{\text{percent}}{100}$$

Before we calculate a percentage, we should understand exactly what a percentage is ? The word percentage comes from the word percent. If you split the word percent into its root words, you see “per” and “cent.” Cent is an old European word with French, Latin, and Italian origins meaning “hundred”. So, percent is translated directly to “per hundred.” If we have 39 percent, we literally have 39 per 100. If it snowed 16 times in the last 100 days, it snowed 16 percent of the time.

Whole numbers, decimals and fractions are converted to percentages. Decimal format is easier to convert into a percentage. Fractions can be converted into decimals and then to percentages.

Converting a decimal to a percentage is as simple as multiplying it by 100. To convert 0.92 to a percent, simply multiple 0.92 by 100.

$0.92 \times 100 = 92\%$

A percentage is an expression of part of the whole. Nothing is represented by 0%, and the whole amount is 100%. Everything else is somewhere in between 0 and 100.

For example, say you have 20 muffins. If you share 12 muffins with your friends, then you have shared 12 out of the 20 muffins ($\frac{12}{20} \times 100\% = 60\%$ shared). If 20 muffins are 100% and you share 60% of them, then 100% - 60% = 40% of the muffins are still left.

Determine the part of the whole :

Given the value for part of the whole and the whole.
OR
Given two parts that make up the whole.

It is important to differentiate what the percentage is "of."
Example : A jar contains 99 gold beads and 51 blue beads. A total of 150 beads are in the jar. In this example, 150 beads makes up a whole jar of beads, i.e. 100%._

1. **Put the two values into a fraction.** The part goes on top of the fraction (numerator), and the whole goes on the bottom (denominator). Therefore the fraction in this case is $\frac{99}{150}$ (part/whole) or $\frac{51}{150}$ (part/whole).

2. **Convert the fraction into a decimal.** Convert $\frac{99}{150}$ into decimal, divide 99 by 150

   $\frac{99}{150} = 0.66$

Converting the decimal into a percent :

Multiply the result obtained in the step above by 100% (per 100 = *per cent*).
0.66 multiplied by 100 equals 66%. Gold beads are 66% of the total beads in the jar.

Converting the percentage into a decimal : Working backward from before, divide the percentage by 100, or you can multiply by 0.01.

$66\% = \frac{66}{100} = 0.66$

Re-word the problem with your new values :

Given in the form of "**X** of **Y** is **Z**." X is the decimal form of your percent, "of" means to multiply, Y is the whole amount, and Z is the answer.

Example: 3% of \$25 is 0.75.
The amount of interest accrued each day on a 3% of loan amount \$25 s 0.75

Discounts :

Discounts are offered on the original price of the item on sale.

1. The discount percent need to be subtracted from the whole, meaning if a discount of 20% is offered on an item priced at \$60 then sale price is 80% of \$60.
   100% - 20% = 80% (whole percent - discount percent).

   80% of \$60 = $\frac{80}{100}$ X 60 = \$48

Markup percentage :

Markup percentage is also referred as gain percentage. Markups are added to the original price of the item before selling it. The greater the markup percentage higher are the profits.

1. The markup percent needs to be added to the whole, meaning if a markup of 20% is added on an item priced at \$60 then sale price is 120% of \$60.
   100% + 20% = 120% (whole percent + markup percent).

   120% of \$60 = $\frac{120}{100}$ X 60 = \$72

Tax percentage :

Taxes are added to the original price of the item before selling it to the customer.

1. The tax percent needs to be added to the whole, meaning if a tax of 5% is added on an item priced at \$60 then sale price is 105% of \$60.
   100% + 5% = 105% (whole percent + tax percent).

   105% of \$60 = $\frac{105}{100}$ X 60 = \$63

Simple Interest :

Simple interest is an easy way of calculating interest on a loan amount.

I = PTR

I = Simple Interest
P = Principle amount of loan
T = Term of the loan in years
R = Rate of interest in decimal form.

Remember final loan amount (A) = P + I

Or

$$I = \frac{PTR}{100}$$

I = Simple Interest
P = Principle amount of loan
T = Term of the loan in years
R = Rate of interest in numerical value only (not in decimal form since the formula has 100 in the denominator).

Remember final loan amount (A) = P + I

Example : Dan took a loan of $1,650 at 7% rate of interest for 5 years from his friend Johnson. Using simple interest find the loan amount to be paid to Johnson at the end of the term.

$$I = \frac{PTR}{100}$$

P = $1,650
R = 7
T = 5 years

$$I = \frac{1650 \text{ X } 7 \text{ X } 5}{100}$$

$$I = \frac{57750}{100}$$

$$= \$577.50$$

Interest to be paid at the end of the loan term is $577.50
Remember the original loan amount $1,650 also need to be paid of.
Amount to be paid of (A) = P + I

A = $1,650 + $577.50
= $2,227.50

Compound interest :

Compound interest is a complex way of calculating interest on a loan amount and the interest on the interest accurred.

$$A = P(1 + \frac{r}{n})^{nt}$$

Where A = Amount at the end of the loan term
Note :It includes principle amount and the interest

P = Principle loan amount
r = Rate of loan or loan percent or interest rate.
Note : Rate needs to be converted to decimal.

n = Number of times interest is compounded per year.
t = Loan term in years.

Remember final loan amount (A) = P + I

OR

$$A = P(1 + \frac{r}{100n})^{nt}$$

Where A = Amount at the end of the loan term
Note :It includes principle amount and the interest

P = Principle loan amount
r = Rate of loan or loan percent or interest rate.
Note : Rate need not be converted to decimal.

n = Number of times interest is compounded per year.
t = Loan term in years.

Remember final loan amount (A) = P + I

Example : Dan took a loan of $49,100 at 14% rate of interest for 2 years compounded annually from his friend Johnson. Using compound interest find the loan amount to be paid to Johnson at the end of the term.

$$A = P\left(1 + \frac{r}{100n}\right)^{nt}$$

P = $49,100
R = 14
T = 2 years
n = 1

$$A = 49100\left(1 + \frac{14}{100}\right)^{2}$$

$$A = 49100\,(1.14)^{2}$$

A = $63,810.36

A = P + I OR I = A - P

Interest I = $63,810.36 - $49,100 = $14,710.36

Interest to be paid at the end of the loan term is $14,710.36
Remember the original loan amount $49,100 also need to be paid of.
Amount to be paid of (A) = P + I

A = $63,810.36

Dan needs to pay $63,810.36 to Johnson at the end of the loan term.

Example : Dan took a loan of $49,100 at 14% rate of interest for 2 years compounded semi annually from his friend Johnson. Using compound interest find the loan amount to be paid to Johnson at the end of the term.

$$A = P(1 + \frac{r}{100n})^{nt}$$

P = $49,100
R = 14
T = 2 years
n = 2

$$A = 49100(1 + \frac{14}{200})^{4}$$

$$A = 49100\ (1.07)^{4}$$

A = $64,360.08

A = P + I OR I = A - P

Interest I = $64,360.08 - $49,100 = $15,260.08

Interest to be paid at the end of the loan term is $15,260.08
Remember the original loan amount $49,100 also need to be paid of.
Amount to be paid of (A) = P + I

A = $64,360.08

Dan needs to pay $64,360.08 to Johnson at the end of the loan term.

Geometry Notes

1. Complementary Angles:

Two angles that add up to 90 degrees are called as Complementary angles.

Example: $\angle X + \angle Y = 90$

$\angle X$ and $\angle Y$ are called complementary angles.

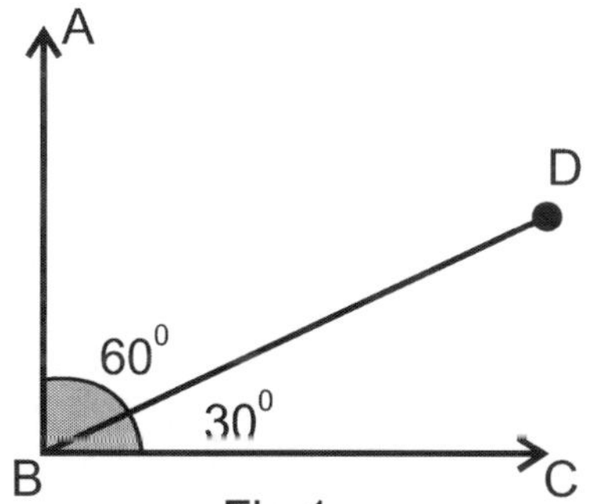

Fig 1

$30^0 + 60^0 = 90^0$
30^0 and 60^0 are complementary angles

2. Supplementary Angles:

Two angles that add up to 180 degrees are called as supplementary angles.

Example: $\angle a + \angle b = 180$

$\angle a$ and $\angle b$ are called complementary angles.

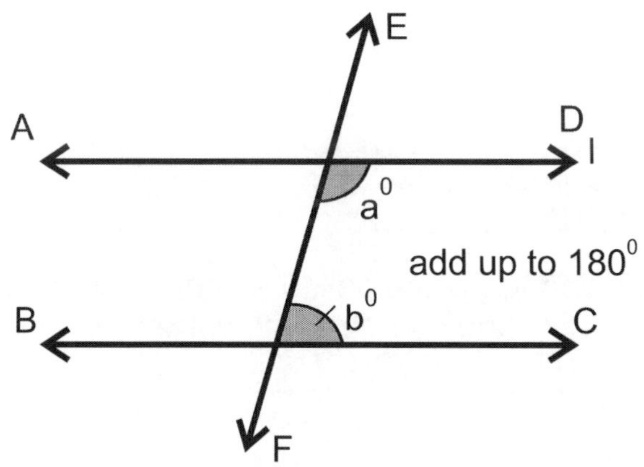

Fig 2

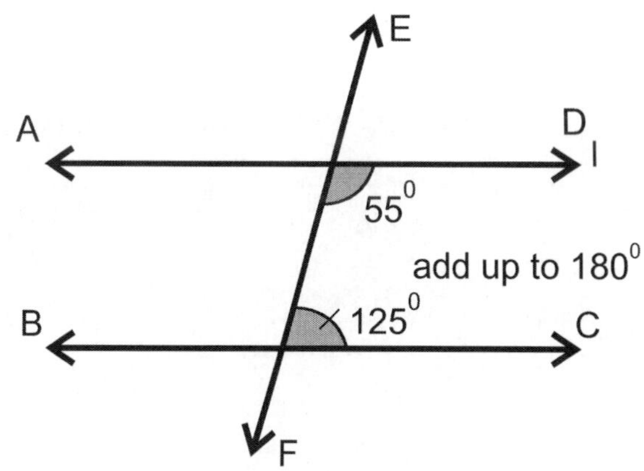

Fig 3

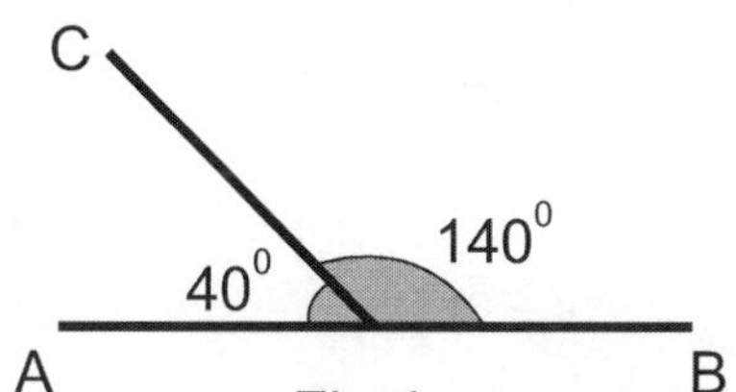

Fig 4

$\angle a + \angle b = 180$; $55^0 + 125^0 = 180^0$

55^0 and 125^0 are called as complementary angles

$\angle a + \angle b = 180$; $40^0 + 140^0 = 180^0$

40^0 and 140^0 are called as complementary angles

3. Vertical Angles:

Vertical angles are pairs of opposite angles made by intersecting lines.
If two angles are vertical, then they are congruent.

Example:

$\angle a$ and $\angle b$ are called vertical angles

$\angle a = \angle b$

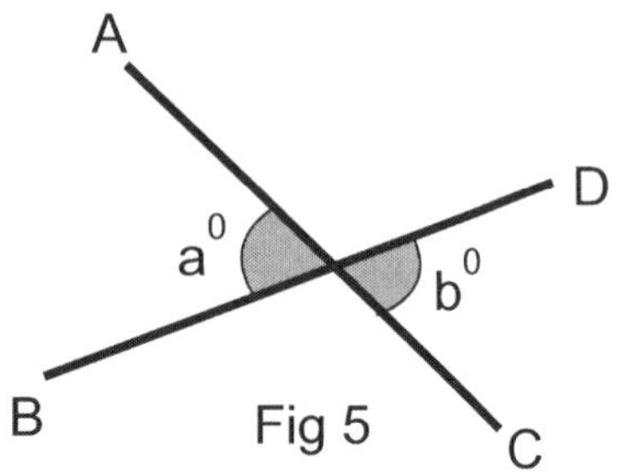

Fig 5

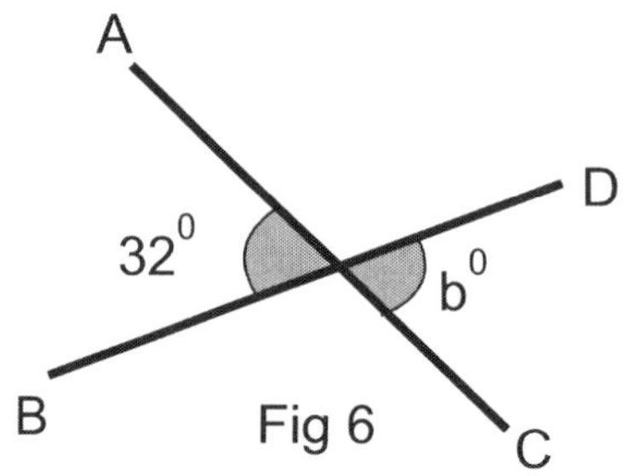

Fig 6

$\angle a$ and $\angle b$ are called vertical angles and vertical angles are equal to each other. In Fig 6 based on the rule of the vertical angles $\angle b = 32^0$

4. Adjacent Angles:

Two angles that have a common side and a common vertex (corner point), and don't overlap are called adjacent angles

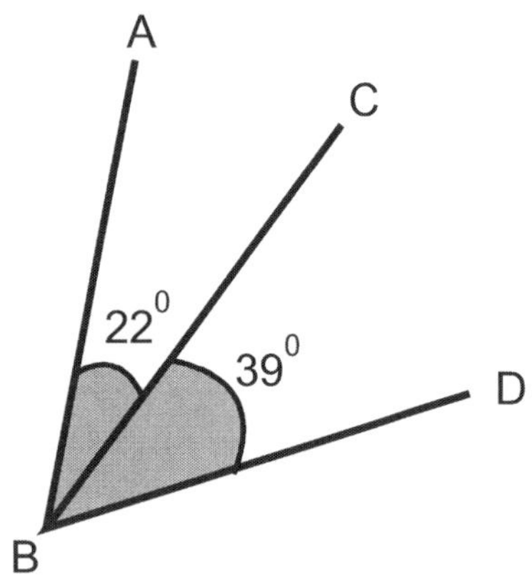

Fig 7

$\angle ABC$ and $\angle CBD$ are called as adjacent angles as they share same vertex B

5. Corresponding Angles

When two lines are crossed by another line (Transversal), the angles in matching corners are called as corresponding angles. A pair of angles each of which is on the same side of one of two lines cut by a transversal and on the same side of the transversal

The angles which occupy the same relative position at each intersection where a straight line crosses two others. If the two lines are parallel, the corresponding angles are equal.

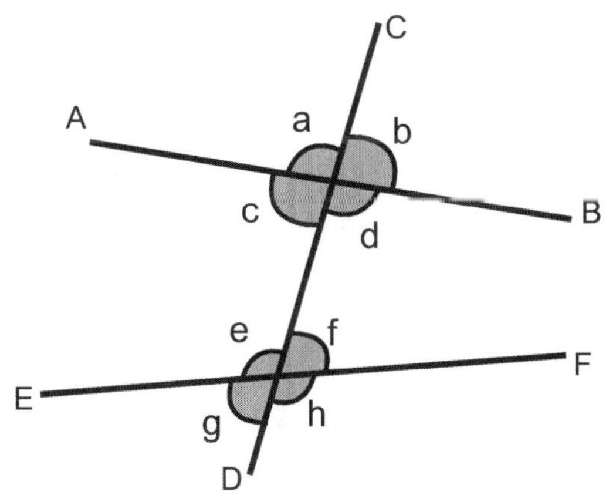

Fig 8

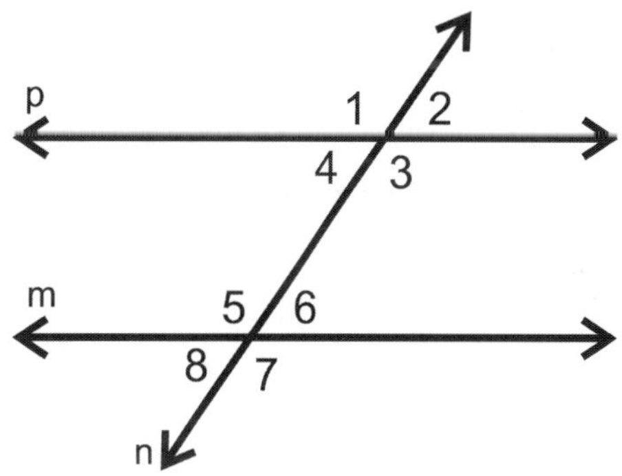

Fig 9

> In Fig 8, $\overline{AB}$ and $\overline{EF}$ **are not parallel**, the corresponding angles ∠a and ∠e ; ∠c and ∠g ; ∠b and ∠f ; ∠d and ∠h are **not equal**.
>
> In Fig 9, Lines p and m **are parallel**, the corresponding angles ∠1 and ∠5 ; ∠4 and ∠8 ; ∠2 and ∠6 ; ∠3 and ∠7 **are equal**.

6. Alternate Angles

Two angles, not adjoining one another, that are formed on opposite sides of a line that intersects two other lines. If the original two lines are parallel, the alternate angles are equal.
one of a pair of angles with different vertices and on opposite sides of a transversal at its intersection with two other lines:

1. Alternate Interior Angles are a pair of angles on the inner side (inside) of each of those two intersected lines but on opposite sides of the transversal. If two parallel lines are cut by a transversal, the alternate interior angles are congruent Examples of Alternate Interior Angles In the figure shown, l is the transversal that cut the pair of lines. Angles 3 and 4 and angles 1 and 2 are alternate interior angles.

2: **Alternate Exterior Angles** are a pair of angles on the outer side (outside) of each of those two intersected lines but on opposite sides of the transversal.

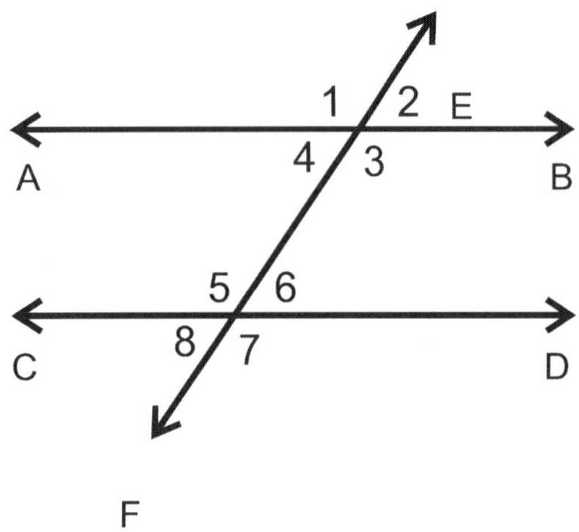

Fig 10

$\angle 3, \angle 4, \angle 5, \angle 6$ Interior Angles

$\angle 1, \angle 2, \angle 7, \angle 8$ Exterior Angles

$\angle 4, \angle 5$; $\angle 3, \angle 6$ } Alternate Interior Angles

$\angle 1, \angle 8$; $\angle 2, \angle 7$ } Alternate Exterior Angles

$\angle 1, \angle 5$; $\angle 2, \angle 6$; $\angle 3, \angle 7$; $\angle 4, \angle 8$ } Corresponding Angles

7. Triangles:
Sum of the angles in any triangle are equal to 180 degrees.

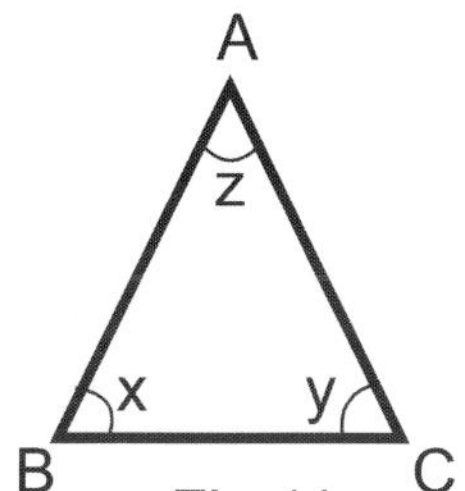

Fig 11

$\angle x + \angle y + \angle z = 180^{0}$

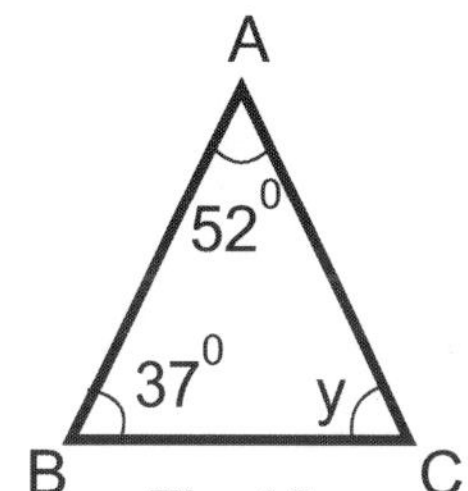

Fig 12

$52^{0} + 37^{0} + \angle y = 180^{0}$

$89^{0} + \angle y = 180^{0}$

$\angle y = 180^{0} - 89^{0}$

$\angle y = 91^{0}$

8. Quadrilaterals: Sum of the angles in any quadrilateral are equal to 360 degrees.

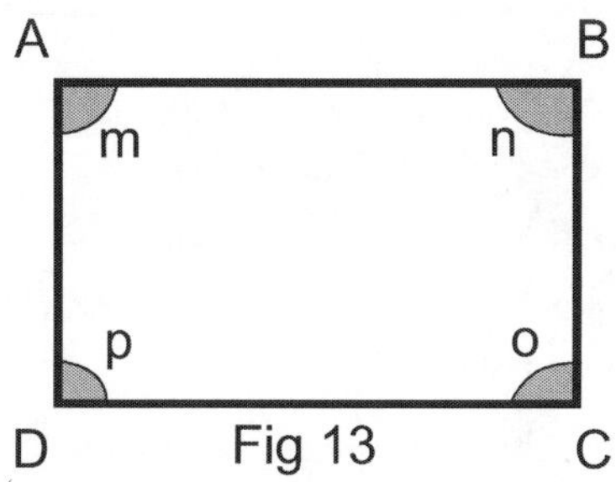

Fig 13

$\angle m + \angle n + \angle o + \angle p = 180^{0}$

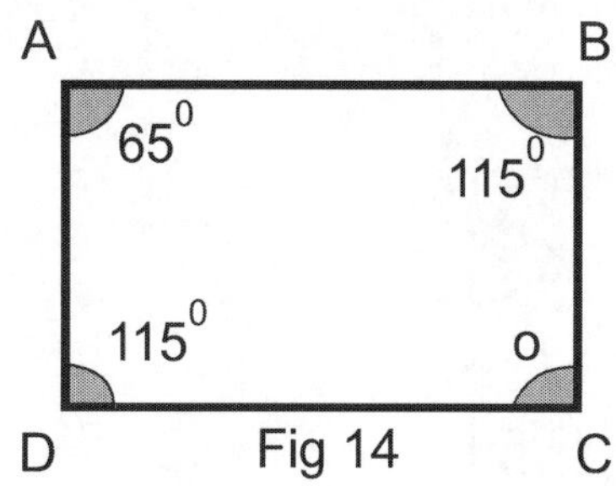

Fig 14

$\angle m + \angle n + \angle o + \angle p = 180^{0}$

$115^{0} + 115^{0} + 65^{0} + \angle p = 180^{0}$

$295^{0} + \angle p = 180^{0}$

$\angle p = 180^{0} - 295^{0}$

$\angle p = 65^{0}$

9. Area of a triangle :

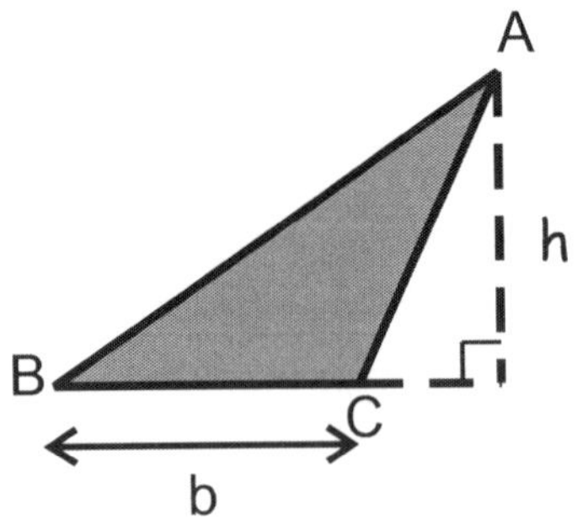

Fig 15

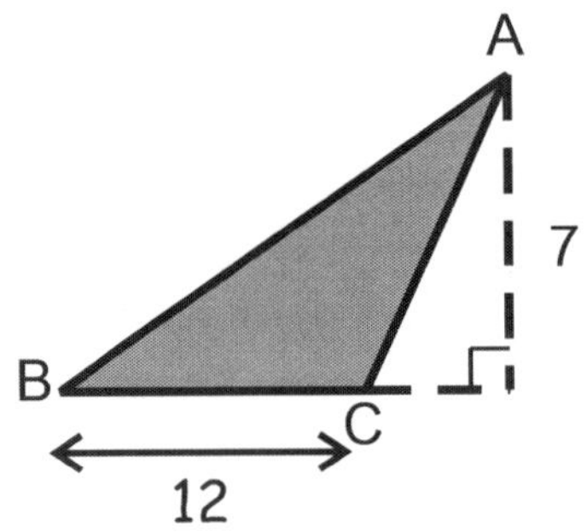

Fig 16

$$A = \frac{1}{2} b h$$

A = Area of the triangle
b = length of the base
h = height

$$A = \frac{1}{2} b h$$

$$A = \frac{1}{2} \times (12)(7)$$

$$= (6)(7)$$

$$= 42 \text{ sq. units}$$

10. Perimeter of a triangle :

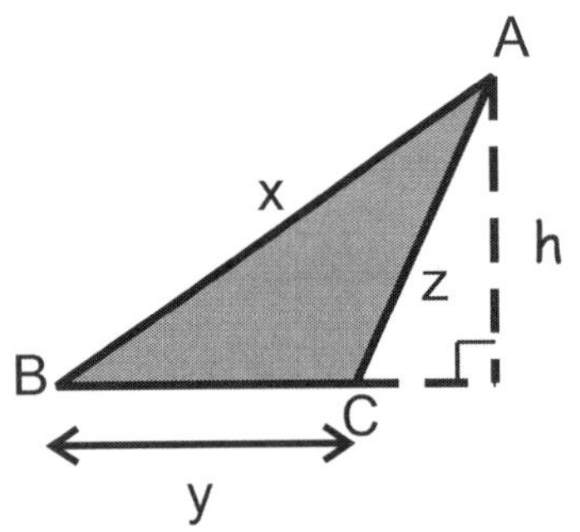

Fig 17

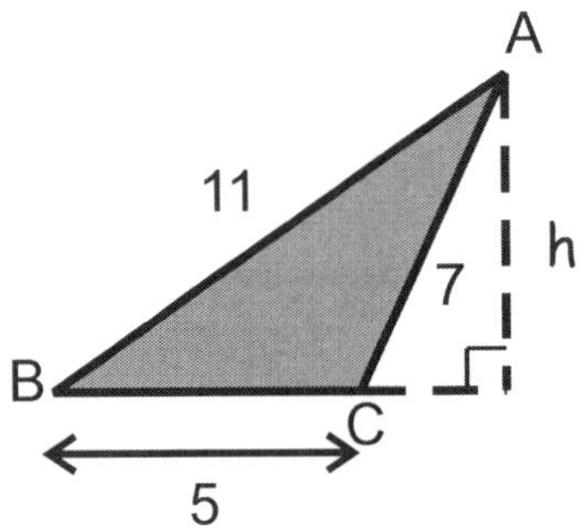

Fig 18

P = x + y + z
P = Perimeter
x, y, z are lengths of the sides of the triangle

P = x + y + z
P = 11 + 5 + 7
P = 23 units

11. Perimeter and Area of a Square

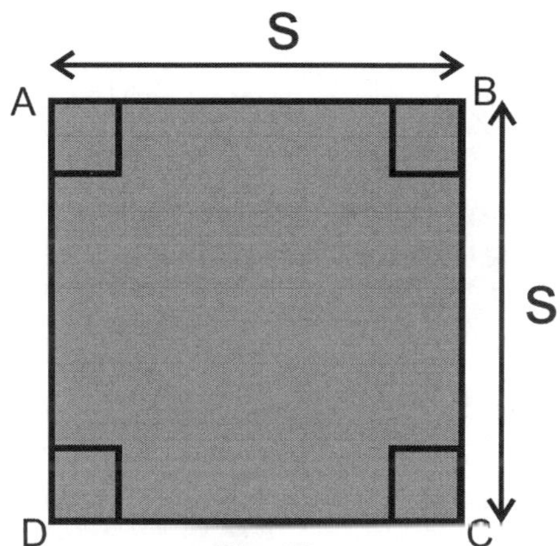

Fig 19

$P = 4s$

$A = s^2$

P = Perimeter
A = Area
s = Length of the side of the square

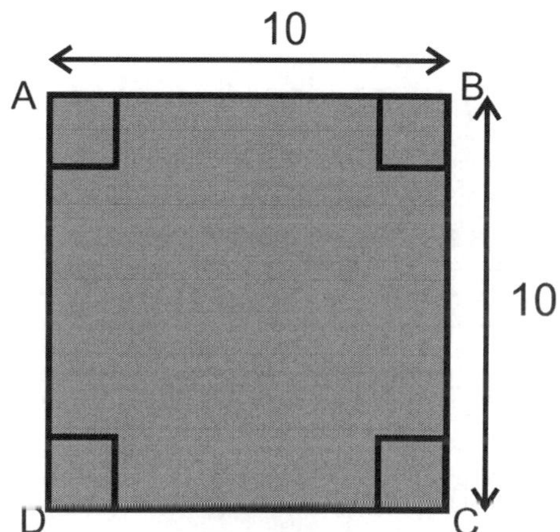

Fig 20

$P = 4s$

$A = s^2$

$P = 4s$

$P = 4 \times 10$

$P = 40$ units

$A = s^2$

$A = 10^2$

$A = 100$ sq.units

12. Perimeter and Area of a Rectangle

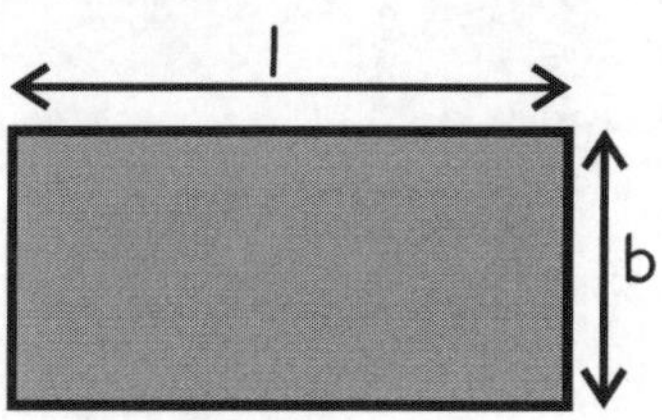

Fig 21

$A = l \times b$

$P = 2(l + b)$

A = Area
P = Perimeter
l = length of the rectangle
b = width of the rectangle

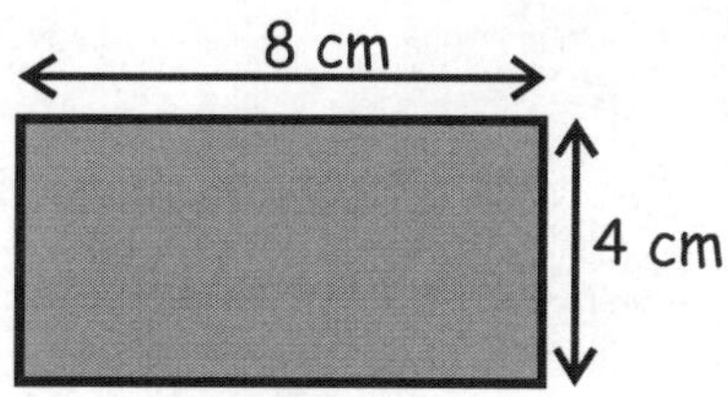

Fig 22

$A = 8 \times 4$

$A = 32$ sq.cm

$P = 2(8 + 4)$

$P = 2(12)$

$P = 24$ cm

13. Area of a Trapezium

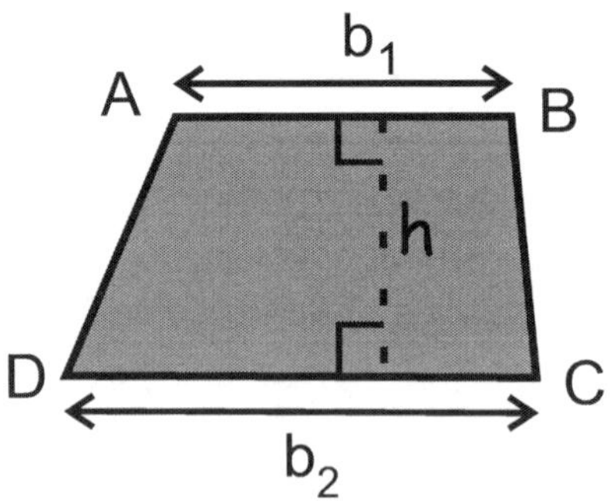

Fig 23

$$A = \frac{1}{2}h(b_1 + b_2)$$

A = Area

b_1, b_2 are the lengths of parallel sides

h = Distance between the parallel sides

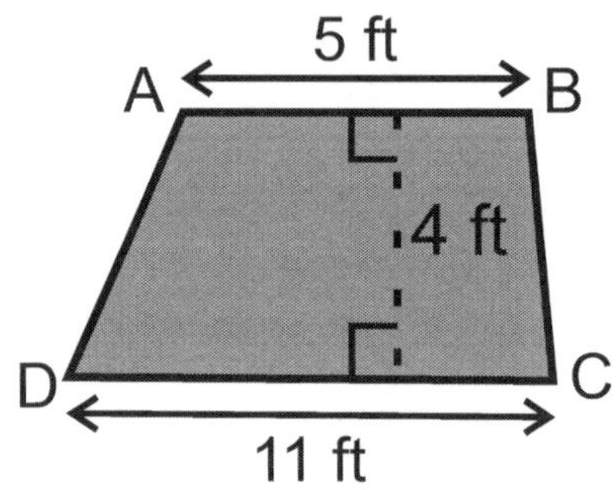

Fig 24

$$A = \frac{1}{2}h(b_1 + b_2)$$

$$A = \frac{1}{2}4(11 + 5)$$

$$A = 2(16)$$

A = 32 sq.ft

14. Perimeter of a Trapezium

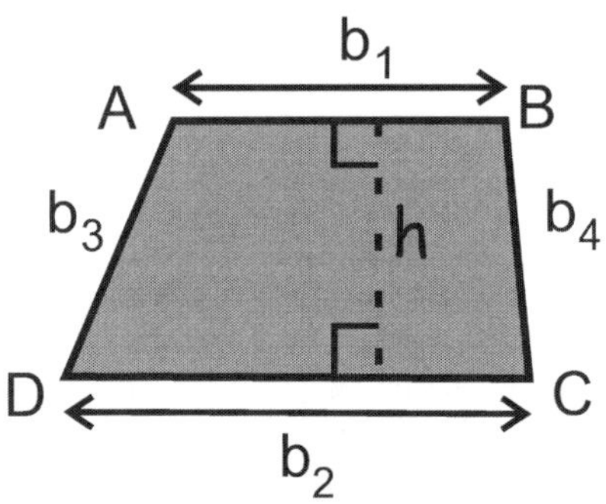

Fig 25

$$P = b_1 + b_2 + b_3 + b_4$$

P = Perimeter

b_1, b_2 are the lengths of parallel sides

b_3, b_4 are the lengths of non parallel sides

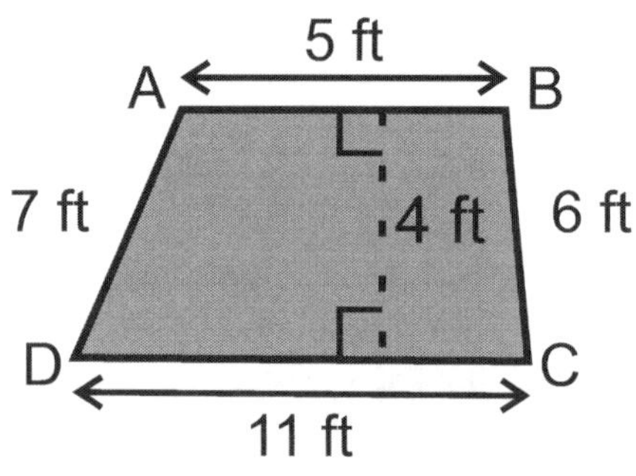

Fig 26

$$P = b_1 + b_2 + b_3 + b_4$$

$$P = 5 + 11 + 7 + 6$$

P = 29 ft

15. Area of a parellelogram

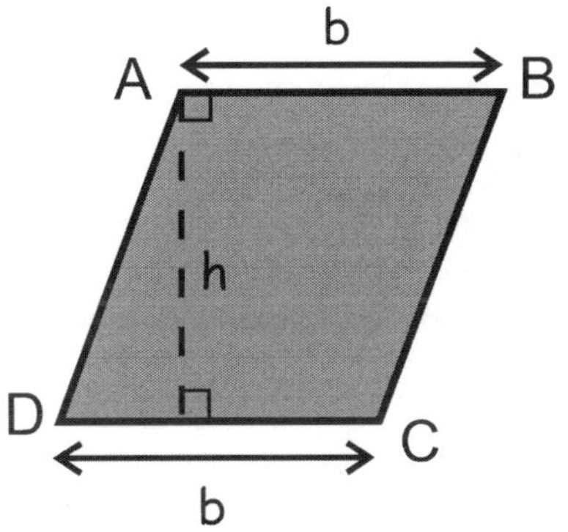

Fig 27

A = bh

A = Area

b = base

h = Height

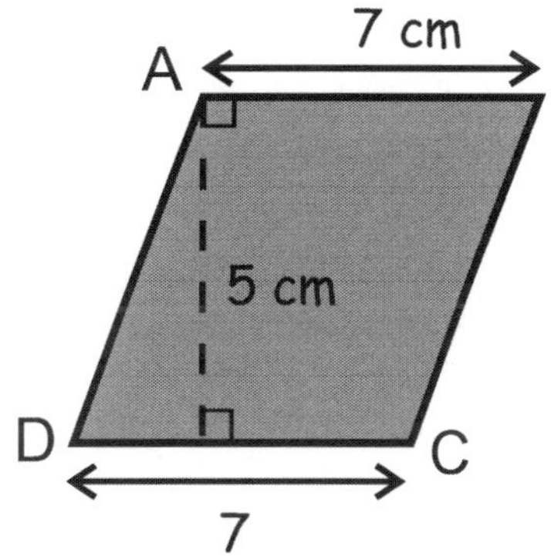

Fig 28

A = bh

A = 5 X 7
A = 45 sq. cm

16. Perimeter of a parellelogram

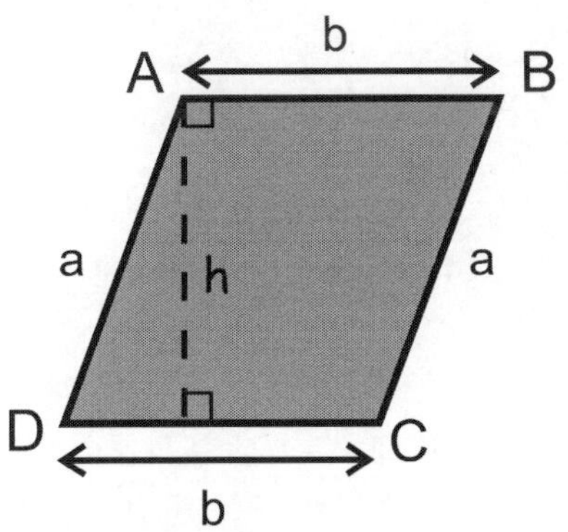

Fig 29

P = a + a + b + b
 = 2(a + b)

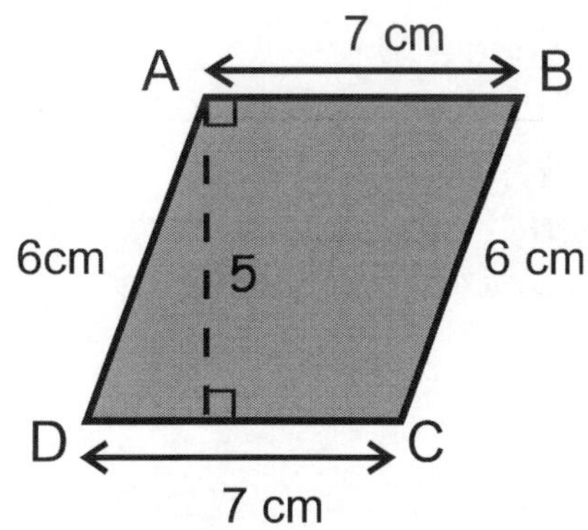

Fig 30

P = 6 + 6 + 7 + 7
 = 2(6 + 7)
 = 2(13)
 = 26 cm

17. Circumference and Area of a Circle

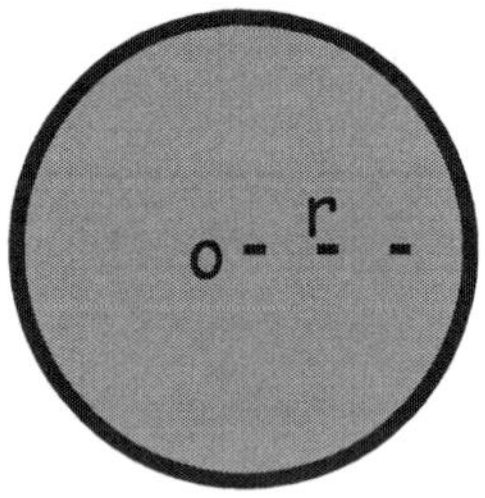

Fig 31

$C = 2\Pi r$

$A = \Pi r^2$

pi

$\Pi = 3.14$

$\Pi = \frac{22}{7}$

C = Circumference of the circle

A = Area of the circle
r = radius

Note : Diameter(d) = 2r

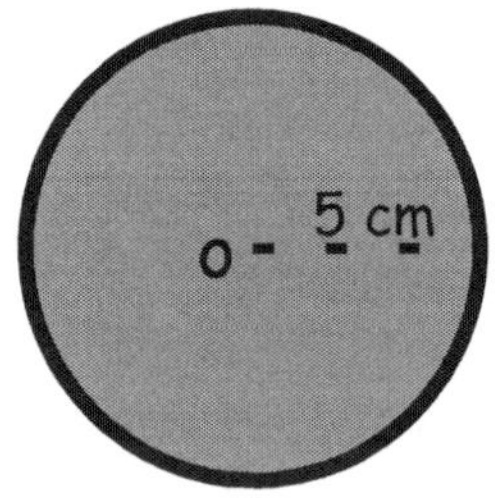

Fig 32

$C = 2\Pi r$

$C = 2 \times 3.14 \times 5$

$C = 10 \times 3.14$

$C = 31.4$ cm

$A = 3.14 \times 5 \times 5$
$A = 3.14 \times 25$
$A = 78.5$ sq.cm

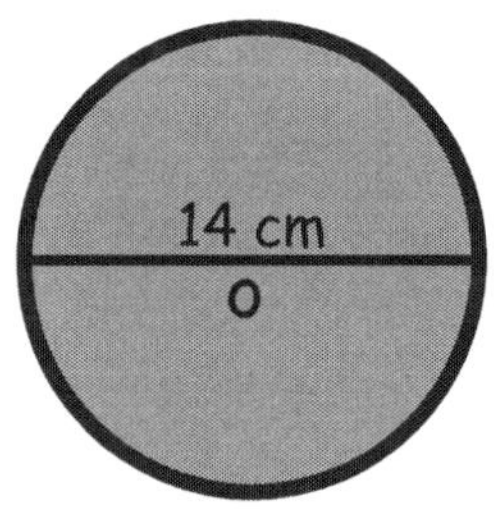

Fig 33

Diameter(d) = 14 cm

Radius $= \frac{14}{2}$

Radius = 7 cm

$C = 2 \times 3.14 \times 7$

$C = 14 \times 3.14$

$C = 43.86$cm

$A = 3.14 \times 7 \times 7$
$A = 3.14 \times 49$
$A = 153.86$ sq.cm

18. Volume of a sphere

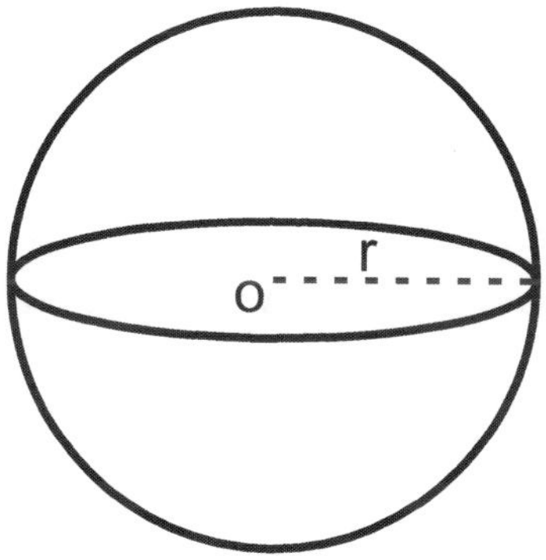

Fig 34

$V = \frac{4}{3} \pi r^3$

V = Volume of the sphere
r = Radius of the sphere

Diameter = Twice the radius

pi

$\pi = 3.14$

$\pi = \frac{22}{7}$

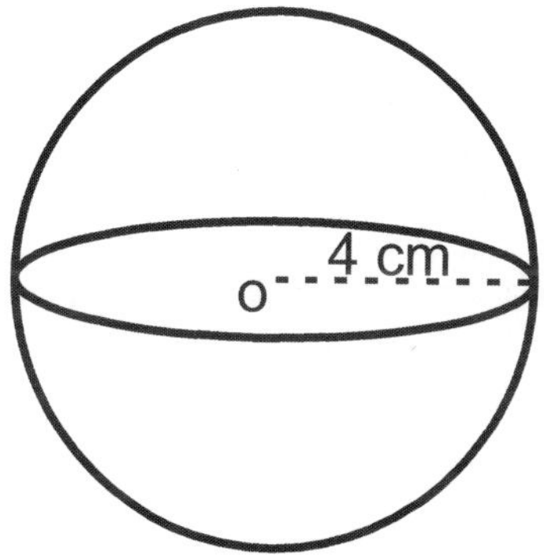

Fig 35

$V = \frac{4}{3} \pi r^3$

$V = \frac{4}{3} \times 3.14 \times 4^3$

$V = 268.08 \text{ cm}^3$

19. Surface area of a sphere

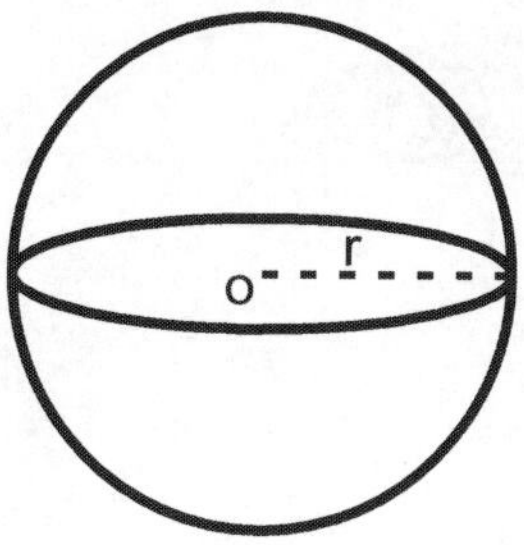

Fig 36

pi

$S = 4\pi r^2$

$\pi = 3.14$

$\pi = \frac{22}{7}$

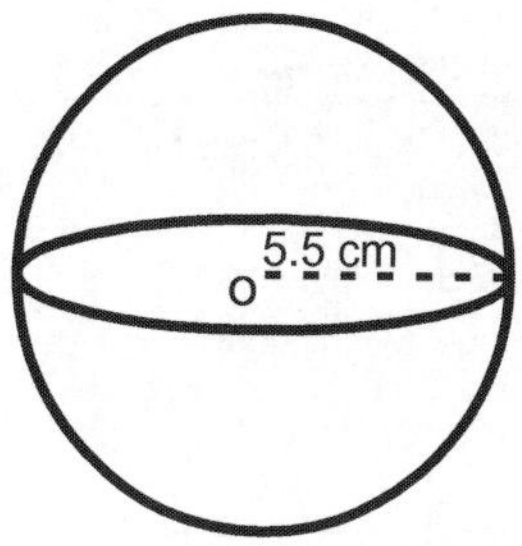

Fig 37

$S = 4\pi r^2$

$S = 4 \times 3.14 \times (5.5)^2$

$S = 379.94 \text{ cm}^2$

20. Volume of a cone

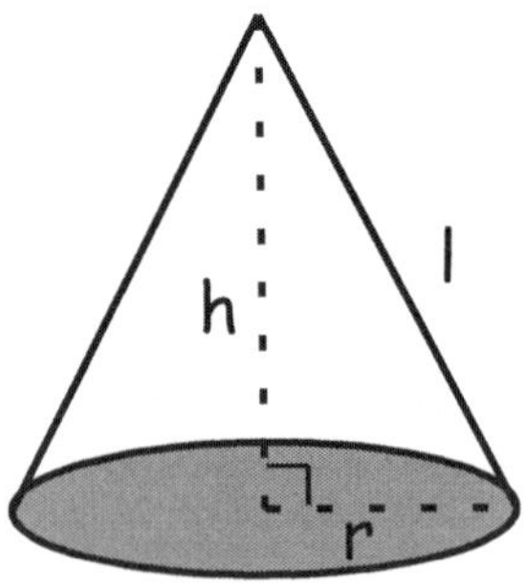

Fig 38

$V = \frac{1}{3}\Pi r^2 h$

V = Volume of the cone
r = Radius
h = Height
l = Slant height

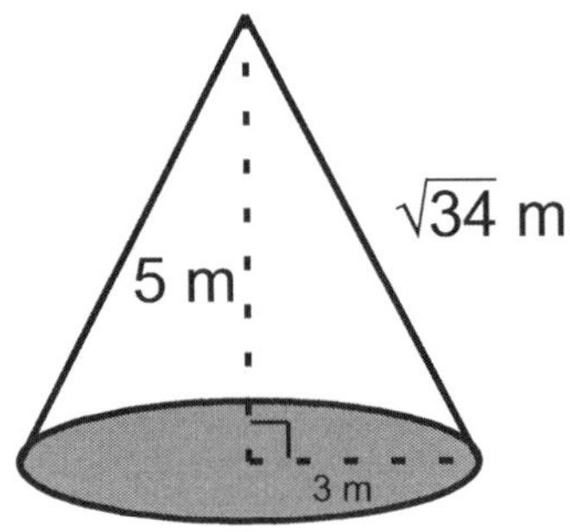

Fig 39

$V = \frac{1}{3}\Pi r^2 h$

$V = \frac{1}{3} X 3.14 X 3^2 X 5$

$V = 47.1239 m^3$

21. Lateral surface area of a cone

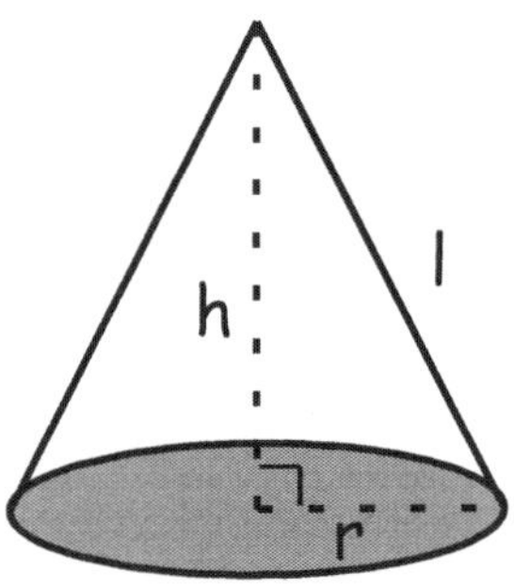

Fig 40

L.S.A $= \Pi rl$

$= \Pi r X \sqrt{r^2 + h^2}$

$l = \sqrt{r^2 + h^2}$

L.S.A = Lateral surface area
l = Slant height of the cone
r = radius
h = height

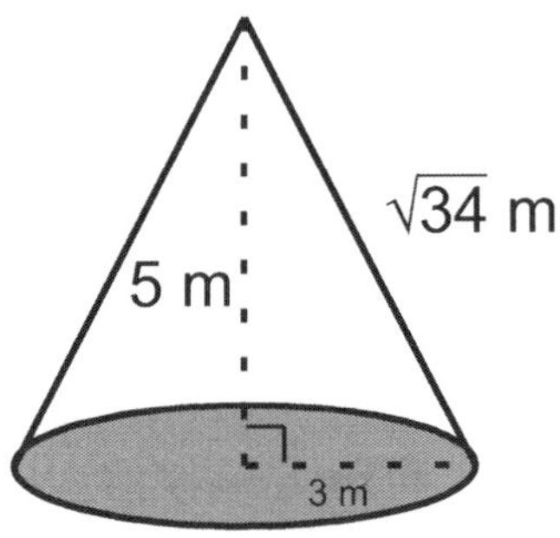

Fig 41

L.S.A $= \Pi rl$

L.S.A = $3.14 X 3 X \sqrt{34}$

$= 54.9554 m^2$

22. Base surface area of a cone

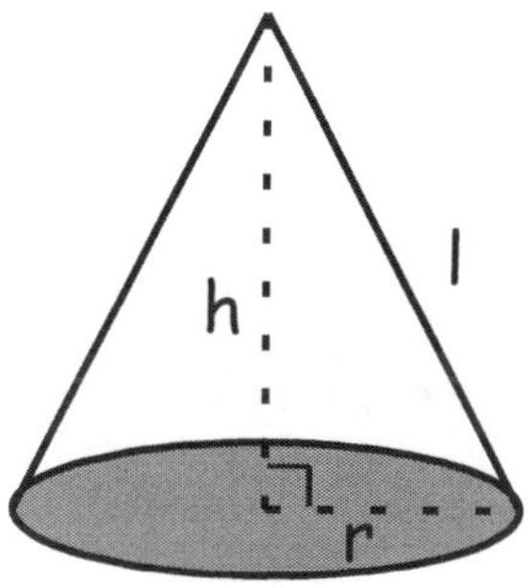

Fig 42

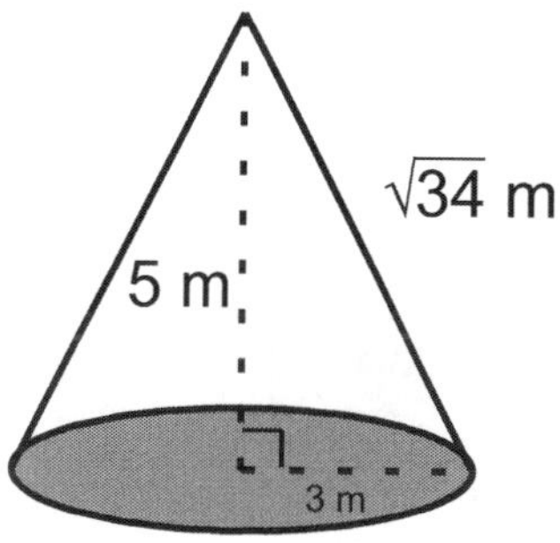

Fig 43

B.S.A $= \prod r^2$

B.S.A = Base surface area
r = radius
h = height

B.S.A $= \prod r^2$

B.S.A = 3.14×3^2

B.S.A = $28.2743\ m^2$

23. Total surface area of a cone

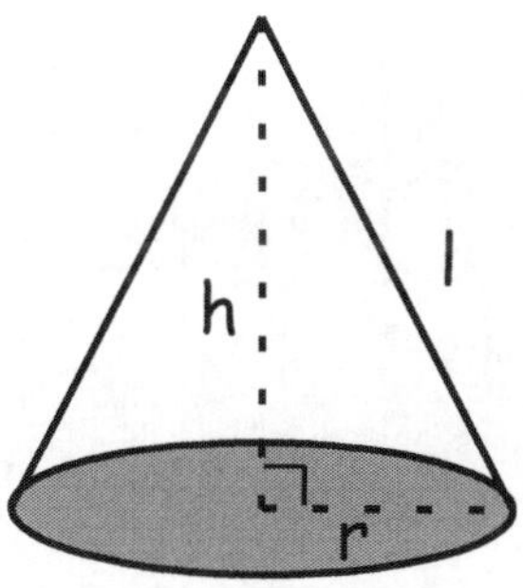

Fig 44

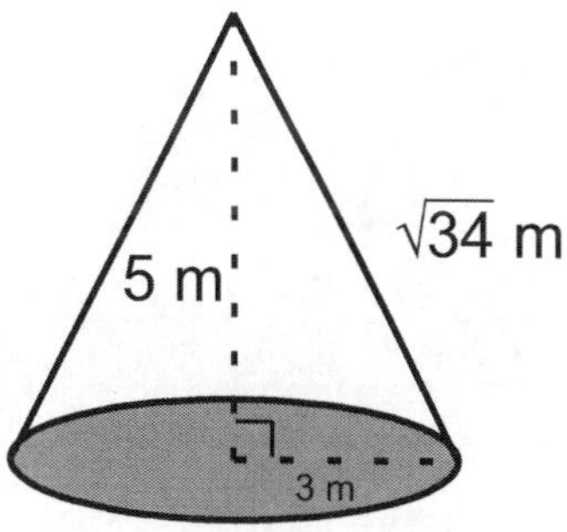

Fig 45

T.S.A = L.S.A + B.S.A

T.S.A = $\prod rl + \prod r^2$

T.S.A = $\prod r(l + r)$

T.S.A = $\prod r(r + \sqrt{r^2 + h^2})$

T.S.A = Total surface area
r = radius
h = height

T.S.A = $\prod r(l + r)$

T.S.A = $3.14 \times 3\ (\sqrt{34} + 3\ )$

T.S.A = $83.2298\ m^2$

24. Volume of a Cylinder

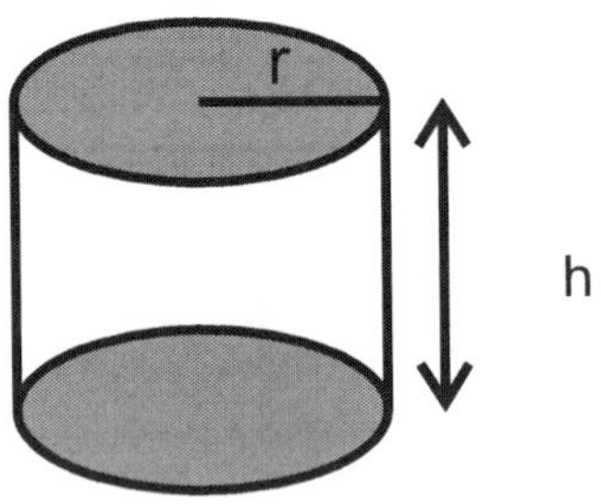

Fig 46

$V = \pi r^2 h$

V = Volume
r = radius
h = height

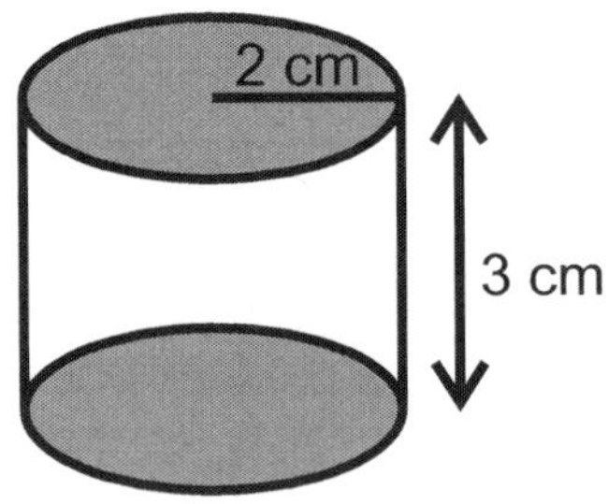

Fig 47

$V = \pi r^2 h$

$V = 3.14 \times 2^2 \times 3$

$V = 37.6991 \text{ cm}^3$

25. Lateral Surface area of a Cylinder

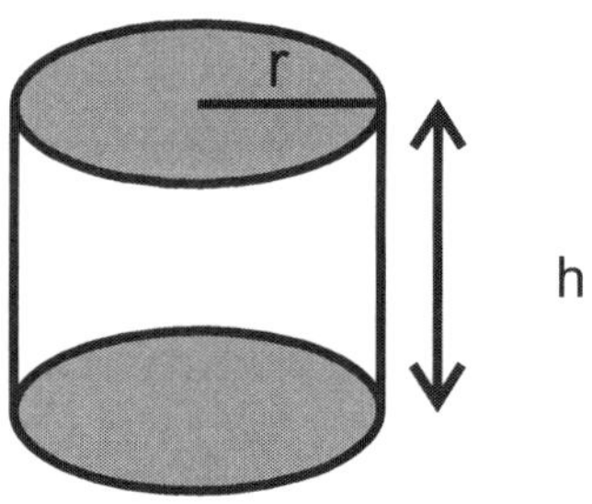

Fig 48

$L.S.A = 2 \pi r h$

L.S.A = Lateral surface area
r = radius
h = height

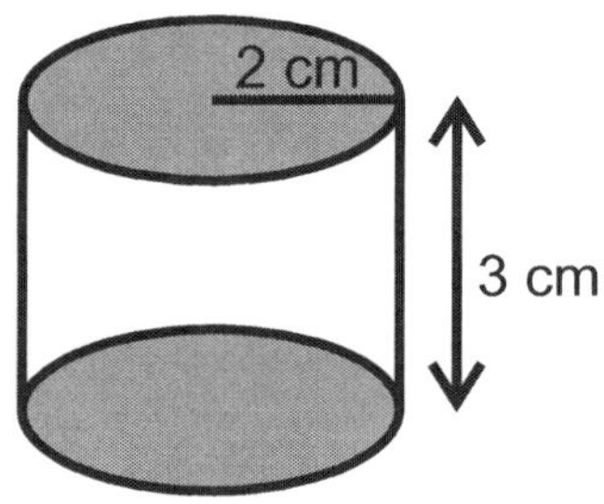

Fig 49

$L.S.A = 2 \pi r h$

$L.S.A = 2 \times 3.14 \times 2 \times 3$

$L.S.A = 37.6991 \text{ cm}^2$

26. Top and bottom Surface area of a Cylinder

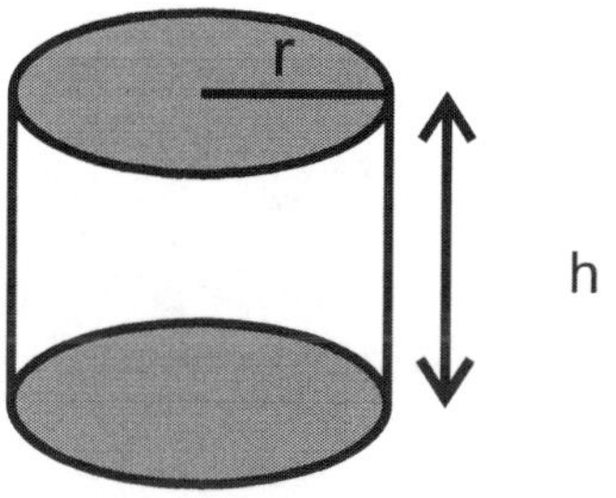

Fig 50

$$B.S.A = \pi r^2$$

B.S.A = Bottom surface area
r = radius
h = height

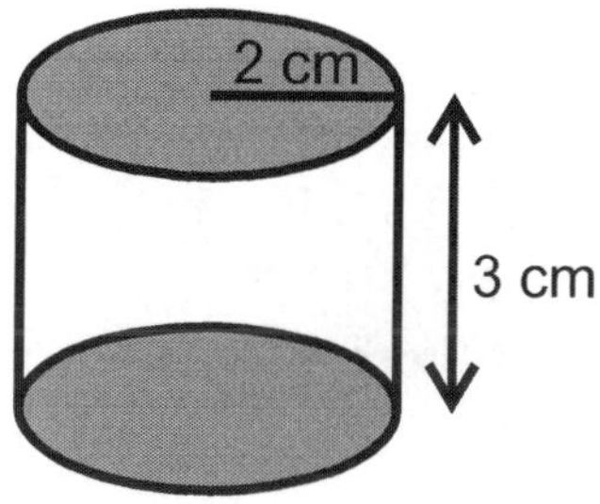

Fig 51

$$B.S.A = \pi r^2$$

$B.S.A = 3.14 \times 2^2$

$B.S.A = 12.5664\ cm^2$

> Top and bottom surface area of the cylinder are equal. While calculating total surface area of the cylinder, remember to add the bottom surface are of the cylinder twice

T.S.A = Total surface area
T.S.A = L.S.A + B.S.A + B.S.A

$T.S.A = 2\pi rh + \pi r^2 + \pi r^2$

$T.S.A = 2\pi rh + 2\pi r^2$

$T.S.A = 2\pi r\,(h + r)$

$T.S.A = 2\pi r\,(h + r)$

$T.S.A = 2 \times 3.14 \times 2\,(3 + 2)$

$T.S.A = 62.8318\ cm^2$

27. Volume of a Cuboid

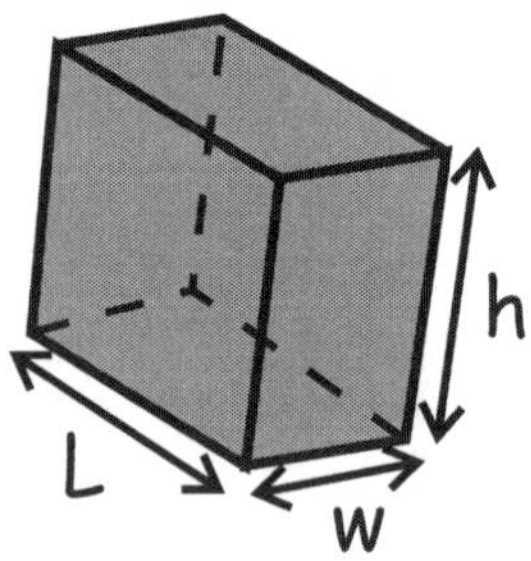

Fig 52

V = lwh
V = Volume
l = Length
w = Width or breadth
h = Height

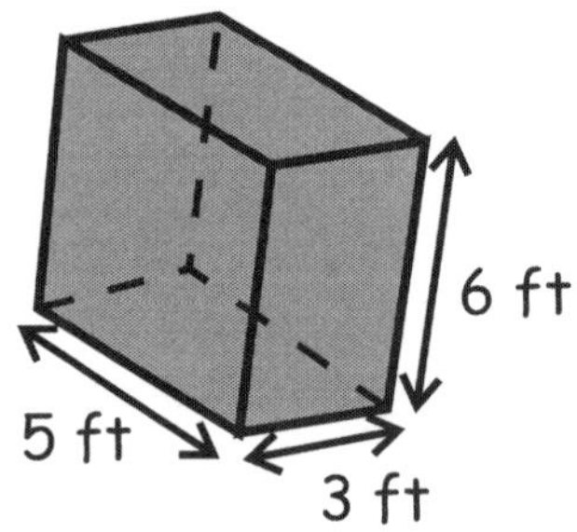

Fig 53

V = lwh

V = 5 X 3 X 6

$V = 90\ ft^3$

28. Surface area of a Cuboid

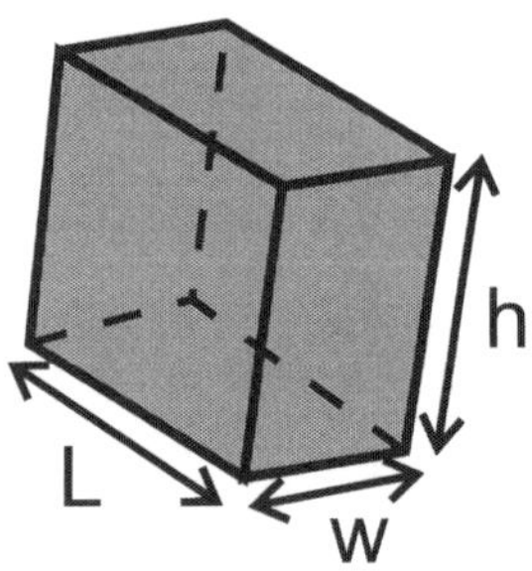

Fig 54

T.S.A = 2(lw + lh + wh)
T.S.A = Total surface area
l = Length
w = Width or breadth
h = Height

L.S.A = 2(lh + wh)
L.S.A = Lateral surface area
l = Length
w = Width or breadth
h = Height

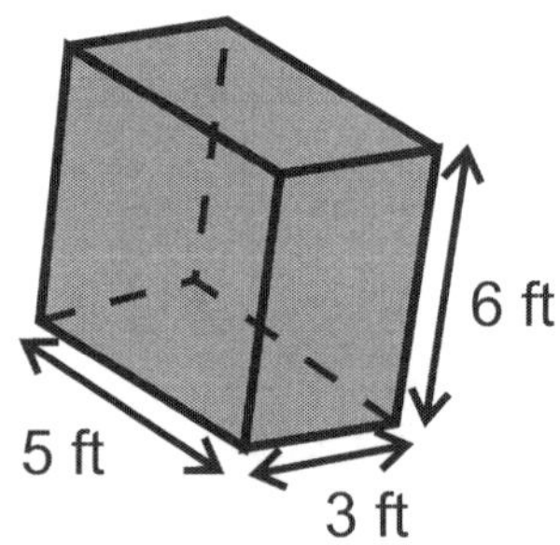

Fig 55

T.S.A = 2(lw + lh + wh)

T.S.A = 2(5X3 + 5X6 + 3X6)

$T.S.A = 126\ ft^2$

L.S.A = 2(lh + wh)

L.S.A = 2(5X6 + 3X6)

$L.S.A = 96\ ft^2$

29. Volume of a Cube

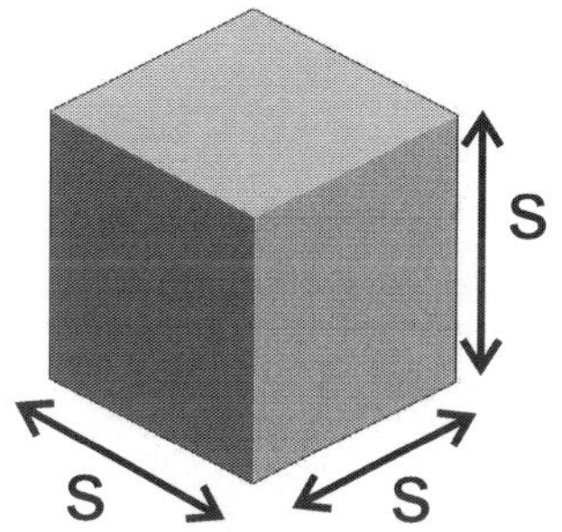

Fig 56

$V = s^3$

V = Volume

s = Side length of the cube

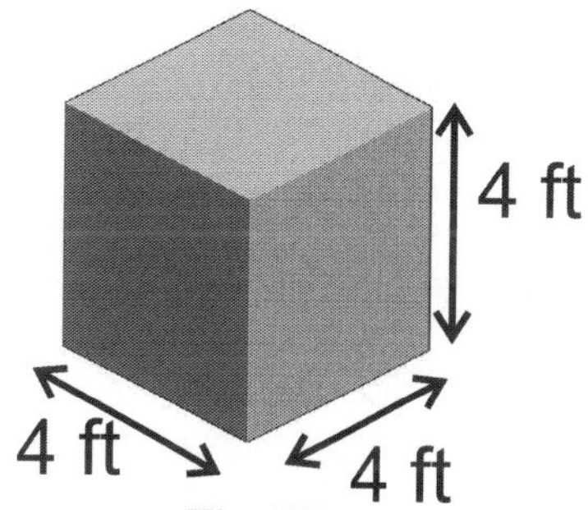

Fig 57

$V = s^3$

$V = 4^3$

$V = 64\ ft^3$

30. Surface area of a Cube

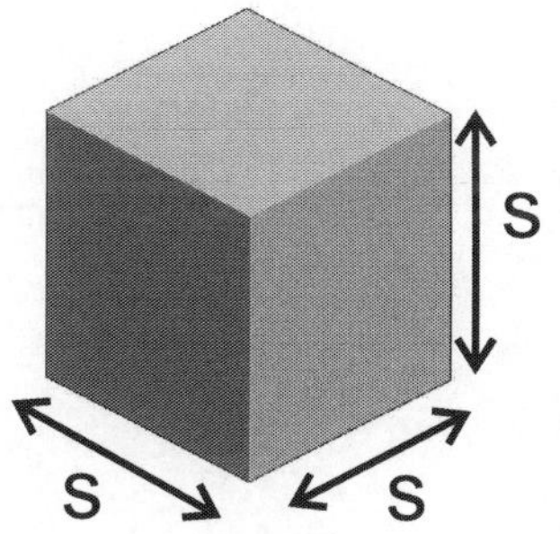

Fig 58

$T.S.A = 6s^2$

T.S.A = Total surface area

s = Side length of the cube

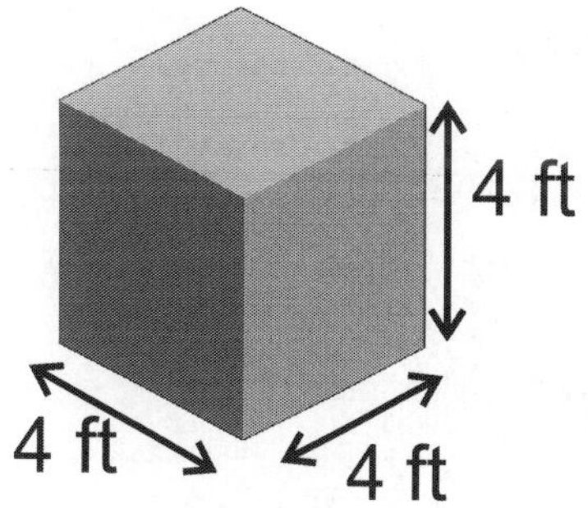

Fig 59

$T.S.A = 6s^2$

$T.S.A = 6 \times 4^2$

$T.S.A = 96\ ft^2$

31. Volume of a square Pyramid

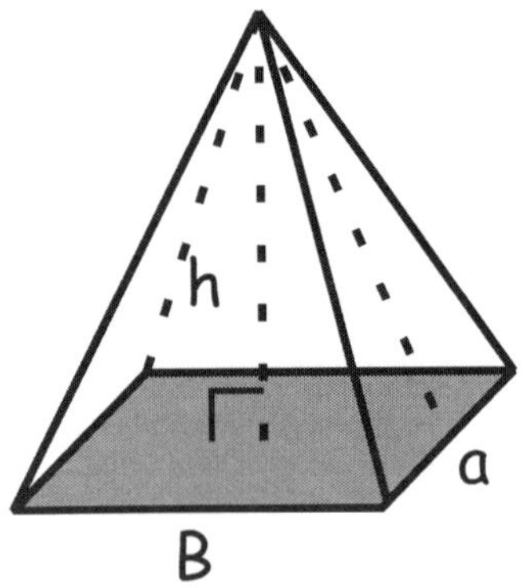

Fig 60

$$V = \frac{1}{3} B h$$

V = Volume
B = Base area
h = Height

$$V = \frac{1}{3} a^2 h$$

V = Volume
a = side length
h = Height

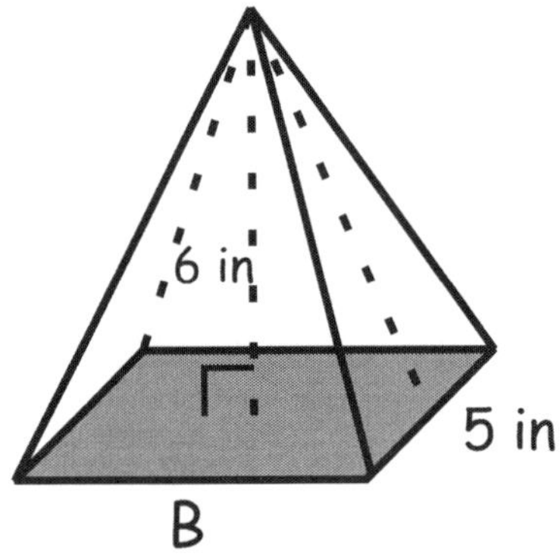

Fig 61

$$V = \frac{1}{3} a^2 h$$

$$V = \frac{1}{3} \times 5^2 \times 6$$

$$V = \frac{1}{3} \times 150$$

$$V = 50 \text{ in}^3$$

32. Lateral surface area of a square Pyramid

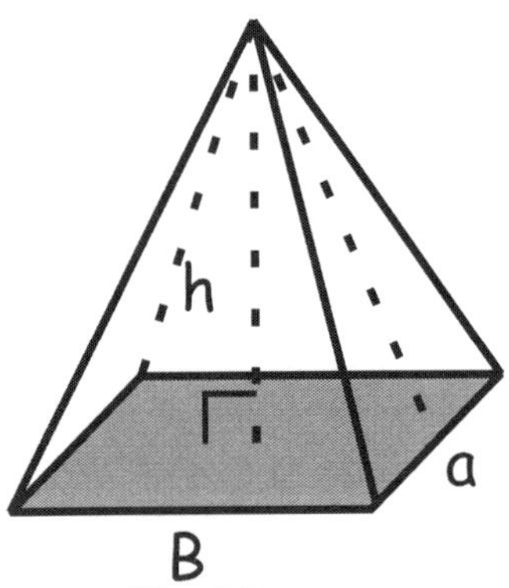

Fig 62

$$S.A = a \times \sqrt{a^2 + 4h^2}$$

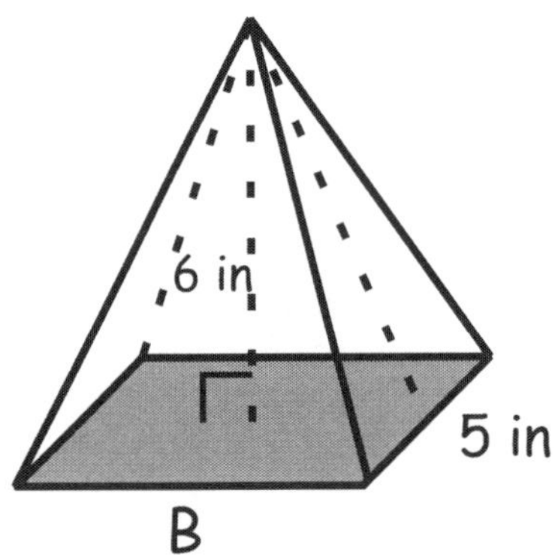

Fig 63

$$S.A = a \times \sqrt{a^2 + 4h^2}$$

$$S.A = 5 \times \sqrt{5^2 + 4 \times 6^2}$$

$$S.A = 65 \text{ in}^2$$

32. Base surface area of a square Pyramid

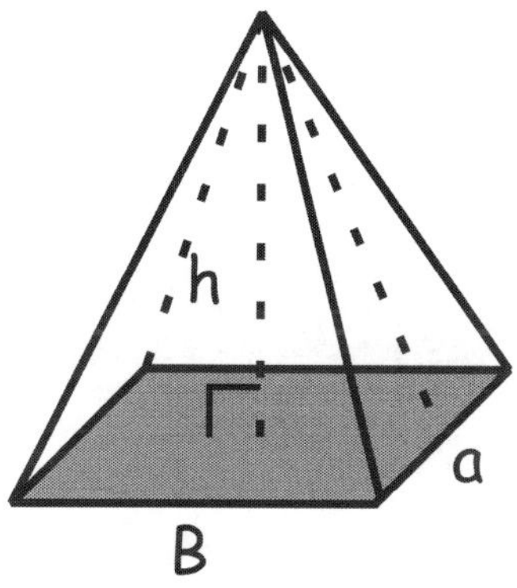

Fig 64

B.S.A = a^2

T.S.A = $a^2 + a \times \sqrt{a^2 + 4h^2}$

B.S.A = Base surface area
T.S.A = Total surface area
T.S.A = B.S.A + L.S.A

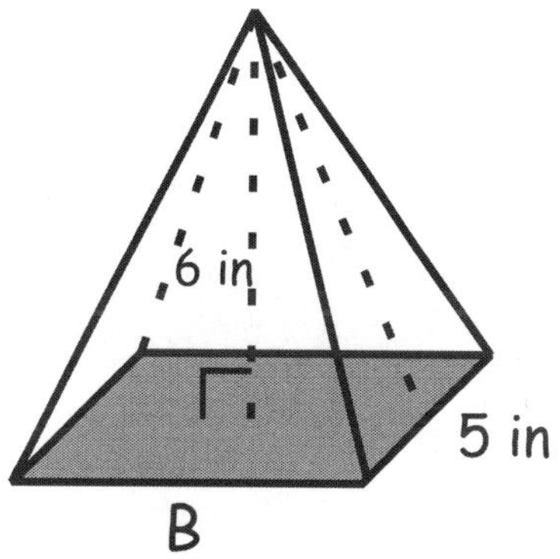

Fig 65

B.S.A = 5 X 5
B.S.A = 25 in^2

T.S.A = $5^2 + 5 \times \sqrt{5^2 + 4 \times 6^2}$

T.S.A = 25 + 65

T.S.A = 90 in^2

32. Volume of a Triangular prism

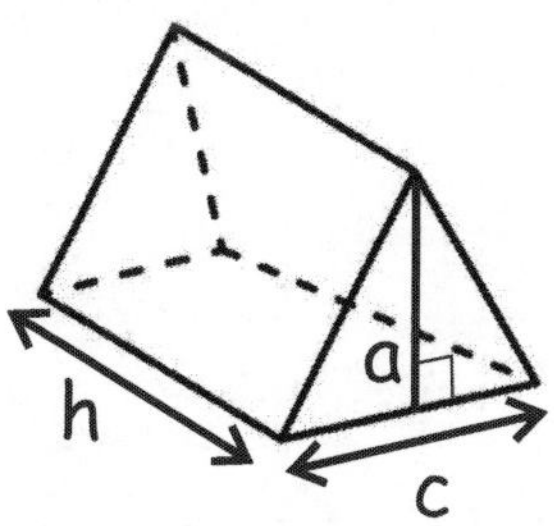

Fig 66

V = Bh

$V = \frac{1}{2}ach$

V = Volume
a = apothem
h = height
c = base length of the triangle

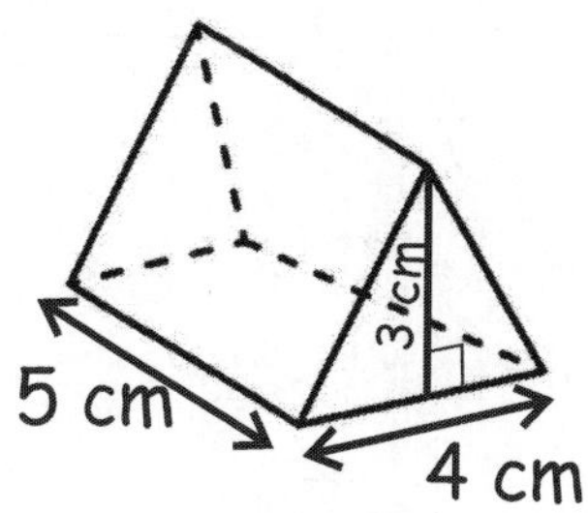

Fig 67

V = Bh

$V = \frac{1}{2}ach$

$V = \frac{1}{2}(3 \times 4 \times 5)$

V = 30 cm^3

32. Volume of a triangular pyramid

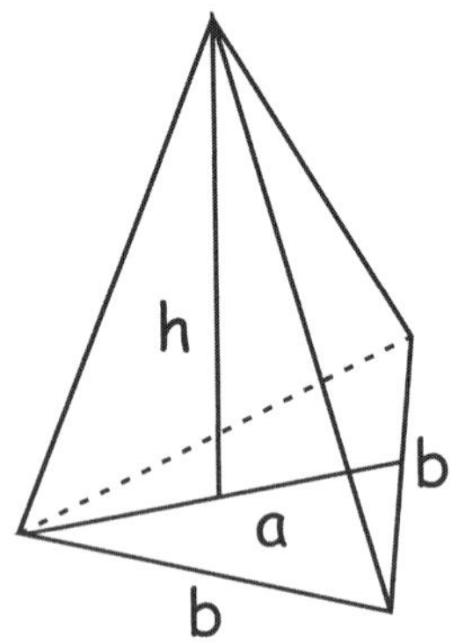

Fig 68

Volume of Triangular pyramid = $\frac{1}{6}$abh

a = Apothem length of the pyramid
b = base length of the pyramid
h = Height of the pyramid

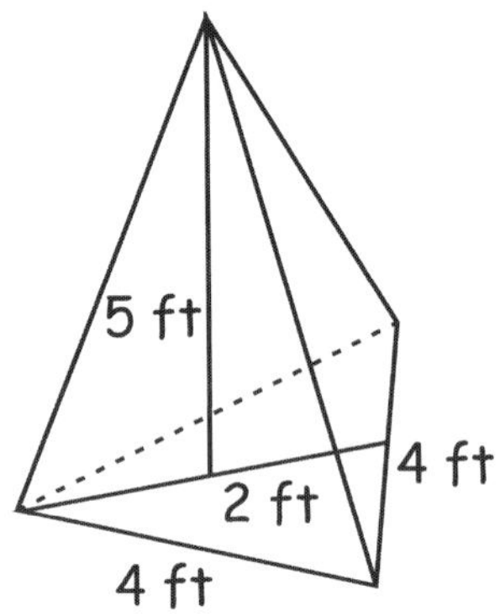

Fig 69

Volume of Triangular pyramid = $\frac{1}{6}$ abh

Volume of Triangular pyramid = $\frac{1}{6}$ (2X4X5)

Volume of Triangular pyramid = $\frac{1}{6}$ (40)

Volume of Triangular pyramid = 6.66 ft^3

33. Volume of a pentagonal pyramid

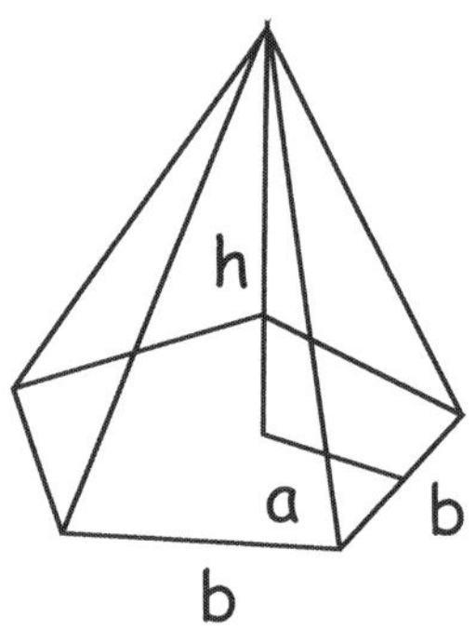

Fig 70

Volume of pentagonal pyramid = $\frac{5}{6}$ abh

a = Apothem length of the pyramid
b = base length of the pyramid
h = Height of the pyramid

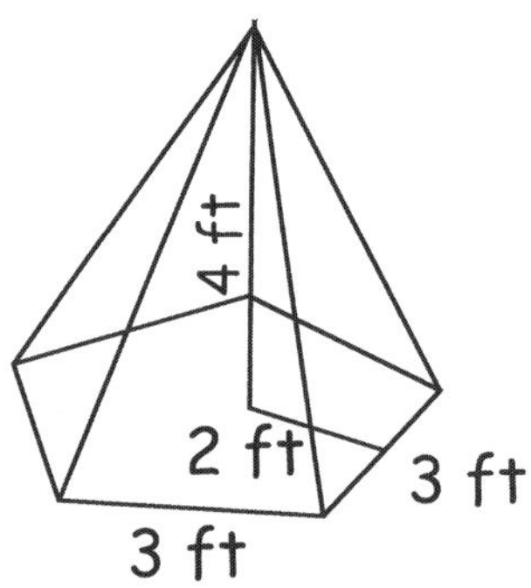

Fig 71

Volume of pentagonal pyramid = $\frac{5}{6}$ abh

Volume of pentagonal pyramid = $\frac{5}{6}$ (2X3X4)

Volume of pentagonal pyramid = $\frac{5}{6}$ (24)

Volume of pentagonal pyramid = 20 ft^3

34. Volume of a hexagonal pyramid

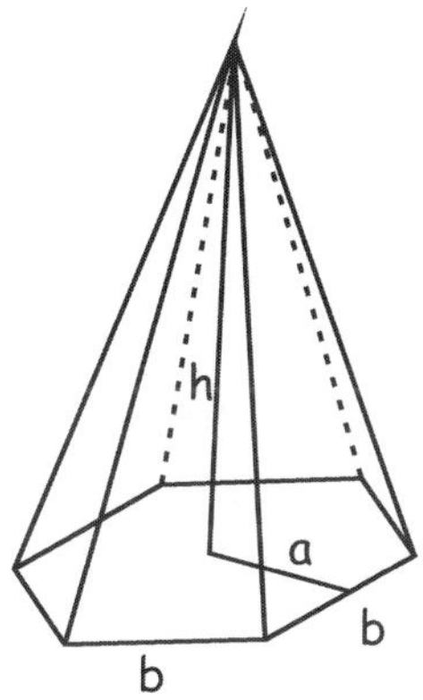

Fig 72

Volume of Hexagonal pyramid = abh

a = Apothem length of the pyramid
b = base length of the pyramid
h = Height of the pyramid

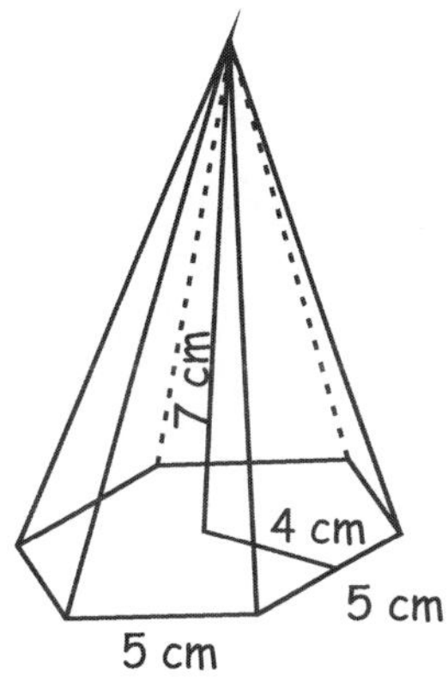

Fig 73

Volume of Hexagonal pyramid = 5X4X7
Volume of Hexagonal pyramid = 140 cm^3

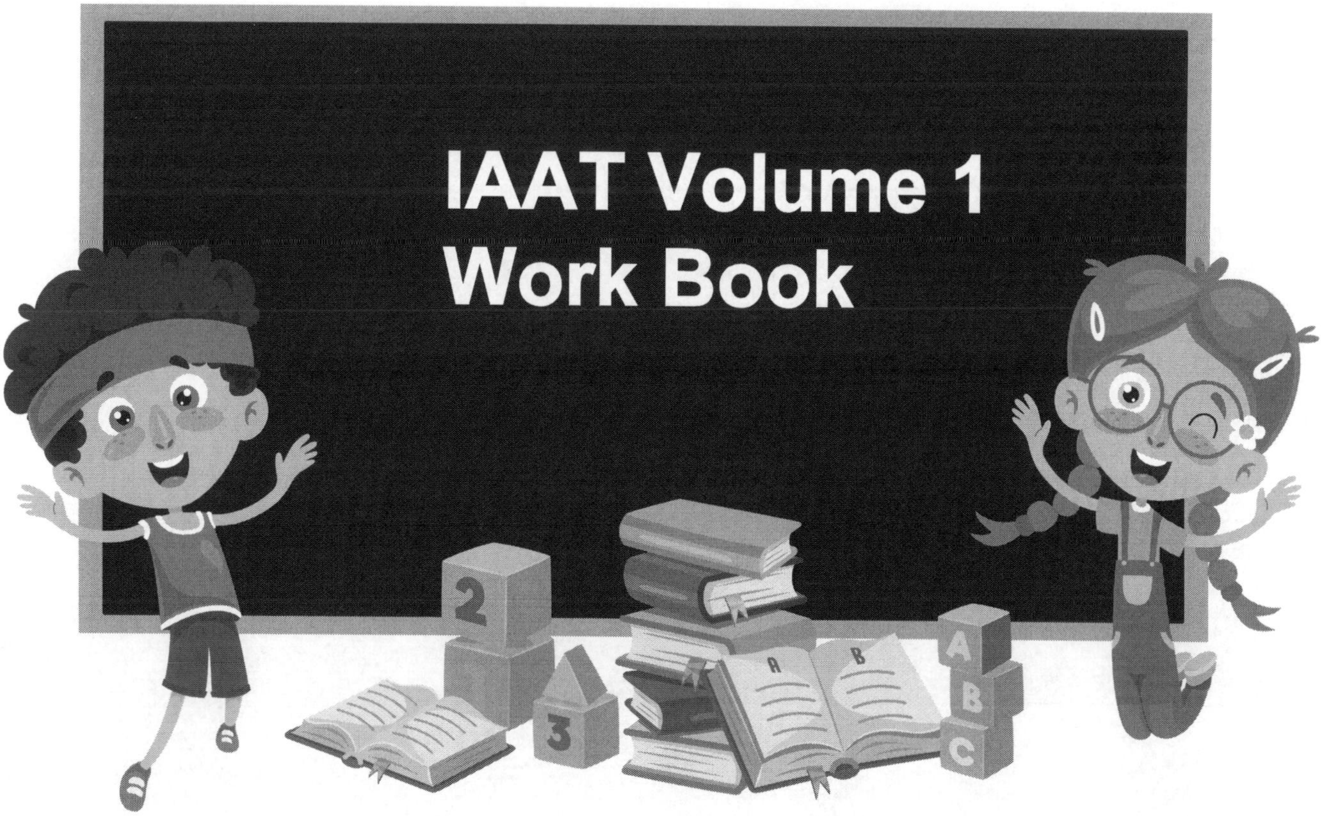
IAAT Volume 1
Work Book

Find the product of the below fractions.

1. $\frac{4}{3} \times \frac{-15}{16}$

(A) $-1\frac{1}{4}$ (B) $\frac{19}{48}$ (C) $\frac{1}{2}$ (D) $1\frac{1}{4}$

2. $-1\frac{5}{8} \times 7\frac{-5}{9}$

(A) $12\frac{5}{18}$ (B) $-16\frac{5}{7}$ (C) $-12\frac{5}{18}$ (D) $16\frac{5}{7}$

3. $15\frac{3}{10} \times -\frac{1}{2}$

(A) $9\frac{3}{20}$ (B) $6\frac{59}{60}$ (C) $7\frac{13}{20}$ (D) $-7\frac{13}{20}$

4. $\frac{1}{4} \times \frac{5}{3}$

(A) $\frac{5}{12}$ (B) $1\frac{1}{12}$ (C) $14\frac{59}{60}$ (D) $\frac{-5}{12}$

Find the product of the below fractions.

5. $-\frac{7}{12} \times \frac{19}{10}$

(A) $-1\frac{13}{120}$ (B) $-1\frac{97}{120}$ (C) $-3\frac{811}{840}$ (D) $1\frac{13}{120}$

6. $\frac{-9}{5} \times -\frac{1}{17}$

(A) $-1\frac{33}{35}$ (B) $\frac{9}{85}$ (C) $-3\frac{31}{560}$ (D) $1\frac{37}{280}$

7. $-3\frac{15}{16} \times 12\frac{1}{2}$

(A) $\frac{12}{65}$ (B) $-49\frac{7}{32}$ (C) $8\frac{9}{16}$ (D) $\frac{-12}{65}$

8. $2\frac{1}{2} \times 2\frac{7}{9}$

(A) $+5\frac{4}{9}$ (B) $-4\frac{17}{18}$ (C) $\frac{5}{18}$ (D) $6\frac{17}{18}$

Find the product of the below fractions.

9. $2\frac{7}{12} \times -\frac{3}{2}$

(A) $-5\frac{3}{8}$ (B) $\frac{29}{40}$ (C) $-3\frac{7}{8}$ (D) $3\frac{7}{8}$

10. $\frac{3}{10} \times -\frac{2}{3}$

(A) $10\frac{31}{90}$ (B) $-\frac{1}{5}$ (C) $-10\frac{31}{90}$ (D) $6\frac{17}{40}$

11. $-1\frac{1}{6} \times \frac{3}{2}$

(A) $-1\frac{3}{4}$ (B) $\frac{1}{3}$ (C) $-9\frac{17}{60}$ (D) $-8\frac{1}{20}$

12. $\frac{1}{4} \times -\frac{13}{12}$

(A) $-\frac{13}{48}$ (B) $-1\frac{409}{624}$ (C) $\frac{13}{48}$ (D) $-2\frac{13}{48}$

Find the product of the below fractions.

13. $\frac{1}{3} \times -\frac{27}{14}$

(A) $\frac{9}{14}$ (B) $-\frac{9}{14}$ (C) $-1\frac{10}{13}$ (D) $1\frac{10}{13}$

14. $2 \times -\frac{3}{7}$

(A) $-12\frac{3}{5}$ (B) $-12\frac{6}{7}$ (C) $-\frac{6}{7}$ (D) $12\frac{3}{5}$

15. $-\frac{21}{13} \times -\frac{9}{5}$

(A) $2\frac{59}{65}$ (B) $-1\frac{503}{520}$ (C) $3\frac{112}{195}$ (D) $-3\frac{27}{65}$

Find each quotient of the below fractions.

1. $-2\frac{1}{13} \div 1\frac{14}{15}$

(A) $1\frac{28}{377}$ (B) $-1\frac{7}{8}$ (C) $-1\frac{28}{377}$ (D) $4\frac{1}{65}$

2. $\frac{8}{13} \div \frac{-4}{3}$

(A) -14 (B) $--\frac{6}{13}$ (C) $2\frac{1}{6}$ (D) $-\frac{32}{39}$

3. $\frac{1}{2} \div 1\frac{4}{5}$

(A) 1 (B) $-1\frac{3}{10}$ (C) $\frac{5}{18}$ (D) $2\frac{3}{5}$

4. $\frac{9}{11} \div \frac{8}{5}$

(A) $1\frac{4}{7}$ (B) $2\frac{23}{55}$ (C) $1\frac{17}{55}$ (D) $\frac{45}{88}$

Find each quotient of the below fractions.

5. $\frac{14}{9} \div -7\frac{1}{3}$

(A) $\frac{7}{33}$ (B) $-11\frac{11}{27}$ (C) $-\frac{7}{33}$ (D) $11\frac{11}{27}$

6. $7\frac{8}{9} \div \frac{2}{5}$

(A) $\frac{4}{15}$ (B) $\frac{2}{9}$ (C) $19\frac{13}{18}$ (D) $1\frac{3}{4}$

7. $\frac{8}{15} \div -6\frac{4}{7}$

(A) $\frac{10}{11}$ (B) $-2\frac{7}{15}$ (C) $-\frac{28}{345}$ (D) $3\frac{53}{105}$

8. $0 \div \frac{-6}{5}$

(A) $4\frac{7}{10}$ (B) 0 (C) $1\frac{11}{12}$ (D) $-1\frac{1}{8}$

Find each quotient of the below fractions.

9. $-2 \div 6\frac{1}{2}$

(A) -13 (B) $-3\frac{1}{2}$ (C) $-\frac{4}{13}$ (D) $\frac{4}{13}$

10. $\frac{-3}{4} \div 4\frac{5}{7}$

(A) $3\frac{15}{28}$ (B) $-3\frac{15}{28}$ (C) $-\frac{7}{44}$ (D) $-6\frac{2}{7}$

11. $-2\frac{5}{7} \div \frac{-23}{15}$

(A) $-1\frac{124}{161}$ (B) $\frac{1}{4}$ (C) $1\frac{124}{161}$ (D) $-4\frac{26}{105}$

12. $\frac{7}{4} \div \frac{5}{13}$

(A) $4\frac{11}{20}$ (B) $-2\frac{11}{12}$ (C) $-\frac{35}{52}$ (D) $-\frac{20}{91}$

Find each quotient of the below fractions.

13. $\frac{5}{6} \div 4\frac{1}{2}$

(A) 7 (B) $-3\frac{3}{4}$ (C) $-\frac{11}{12}$ (D) $\frac{5}{27}$

14. $\frac{4}{5} \div \frac{-5}{4}$

(A) $2\frac{1}{20}$ (B) $-1\frac{2}{3}$ (C) $1\frac{9}{16}$ (D) $-\frac{16}{25}$

15. $-1\frac{2}{9} \div 1\frac{1}{14}$

(A) $-1\frac{11}{3}$ (B) $-2\frac{37}{126}$ (C) $-2\frac{1}{6}$ (D) $-1\frac{19}{135}$

Simplify the below fractions.

1. $2\frac{13}{48} - \frac{17}{42}$

(A) $3\frac{317}{560}$ (B) $1\frac{97}{112}$ (C) $24\frac{425}{1008}$ (D) $4\frac{2731}{3024}$

2. $\frac{3}{2} - 1\frac{1}{7}$

(A) $22\frac{83}{154}$ (B) $\frac{5}{14}$ (C) $1\frac{27}{56}$ (D) $14\frac{1}{42}$

3. $33\frac{41}{42} - \frac{11}{24}$

(A) $50\frac{347}{2184}$ (B) $23\frac{1529}{2520}$ (C) $33\frac{29}{56}$ (D) $32\frac{43}{56}$

4. $\frac{28}{17} - \frac{27}{26}$

(A) $\frac{269}{442}$ (B) $\frac{610}{1547}$ (C) $9\frac{269}{442}$ (D) $23\frac{1393}{2652}$

Simplify the below fractions.

5. $\frac{51}{26} - \frac{48}{25}$

(A) $\frac{27}{650}$ (B) $3\frac{27}{650}$ (C) $10\frac{577}{650}$ (D) $12\frac{25}{26}$

6. $35 - 6\frac{19}{46}$

(A) $29\frac{893}{2070}$ (B) $53\frac{347}{1610}$ (C) $27\frac{479}{2070}$ (D) $28\frac{27}{46}$

7. $\frac{1}{5} - \frac{1}{6}$

(A) $\frac{1189}{1470}$ (B) $1\frac{7}{10}$ (C) $\frac{1}{30}$ (D) $7\frac{11}{15}$

8. $7\frac{1}{17} - \frac{42}{43}$

(A) $6\frac{2383}{3519}$ (B) $6\frac{60}{731}$ (C) $5\frac{64}{1955}$ (D) $11\frac{10142}{18377}$

Simplify the below fractions.

9. $22\frac{8}{17} - 7\frac{9}{34}$

(A) $18\frac{579}{748}$ (B) $16\frac{25}{204}$ (C) $22\frac{789}{1190}$ (D) $15\frac{7}{34}$

10. $\frac{14}{9} - \frac{5}{22}$

(A) $3\frac{65}{198}$ (B) $1\frac{65}{198}$ (C) $2\frac{4589}{6138}$ (D) $11\frac{755}{792}$

11. $16\frac{11}{17} - \frac{9}{7}$

(A) $17\frac{741}{2380}$ (B) $15\frac{43}{119}$ (C) $15\frac{1650}{2261}$ (D) $13\frac{709}{833}$

12. $14\frac{13}{20} - \frac{19}{11}$

(A) $13\frac{257}{660}$ (B) $12\frac{203}{220}$ (C) $10\frac{204}{385}$ (D) $14\frac{27}{220}$

Simplify the below fractions.

13. $15\frac{11}{13} - 5\frac{28}{29}$

(A) $10\frac{332}{377}$ (B) $9\frac{332}{377}$ (C) $23\frac{332}{377}$ (D) $10\frac{3163}{17719}$

14. $46 - \frac{46}{31}$

(A) $46\frac{17}{93}$ (B) $45\frac{458}{527}$ (C) $45\frac{631}{682}$ (D) $44\frac{16}{31}$

15. $16\frac{7}{33} - \frac{35}{41}$

(A) $15\frac{6101}{9471}$ (B) $15\frac{485}{1353}$ (C) $15\frac{7658}{17589}$ (D) $5\frac{3639}{7216}$

Evaluate the following.

1. $9.7 + (-29.7) + (-2.2)$

(A) -57.1 (B) -22.2 (C) 51.7 (D) 46.1

2. $91 - (-93.4) + 39.9$

(A) 315.5 (B) 267.3 (C) 224.3 (D) 222.3

3. $74.3 + (-90.4) + 75.1$

(A) 59 (B) 51.4 (C) -35.2 (D) 142.86

4. $93.6 - 39.9 - 76.52$

(A) -22.82 (B) -89.32 (C) 18.18 (D) -111.12

5. $(-7.6) - 13.2 - (-64)$

(A) 29.1 (B) -21.085 (C) 29.37 (D) 43.2

Evaluate the following.

6. $(-77.87) + (-49.7) - (-12.821)$

(A) −208.749 (B) −180.349 (C) −141.049 (D) −114.749

7. $(-45.5) - (-50.65) + (-19)$

(A) −61.55 (B) −107.85 (C) 65.45 (D) −13.85

8. $63 - (-29.8) + (-54.1)$

(A) 111 (B) 31.8 (C) 38.7 (D) 54.6

9. $(-50.8) - (-91.9) - 11.8$

(A) −23.2 (B) 79.6 (C) 78.2 (D) 29.3

10. $(-70.7) + 5.7 + (-94.53)$

(A) −102.03 (B) −177.03 (C) −159.53 (D) −156.23

Evaluate the following.

11. $(-66.8) - 60.675 - (-28.337)$

(A) -91.638 (B) -106.838 (C) -99.138 (D) -183.738

12. $48.3 + 51.26 - 11.19$

(A) 88.37 (B) 30.77 (C) 33.67 (D) 179.67

13. $56.8 + (-28.9) - (-43.3)$

(A) -0.8 (B) 131 (C) 71.2 (D) -19.5

14. $(-13.7) - (-41.3) - (-48.7)$

(A) 100.6 (B) 31.4 (C) 28 (D) 76.3

15. $(-99.6) + (-56) + 79.4$

(A) -61 (B) -95.4 (C) -76.2 (D) -149.64

Find the product of the below.

1. (-5.9)(0.8)

(A) -4.72 (B) 2.78 (C) 4.72 (D) -5.1

2. (-5.3)(1.1)

(A) -5.83 (B) -5.33 (C) -2.73 (D) -4.2

3. (-3.9)(-5.2)

(A) -20.28 (B) -9.1 (C) 23.17 (D) 20.28

4. (3.7)(-9.965)

(A) 36.8705 (B) -26.9095 (C) -36.8705 (D) -6.265

5. (-7.6)(2.6)

(A) -13.66 (B) -26.96 (C) -19.76 (D) 19.76

Find the product of the below.

6. $(0.8)(-3.6)$

(A) 2.88 (B) −2.8 (C) −8.88 (D) −2.88

7. $(-7.437)(-7.1)$

(A) 52.8027 (B) −52.8027 (C) 46.3027 (D) −14.537

8. $(-8.05)(-1.3)$

(A) 10.465 (B) −48.5 (C) 48.5 (D) 14.265

9. $(-9.1)(1.9)$

(A) −17.29 (B) 17.29 (C) −7.2 (D) −24.899

10. $(-5.8)(7.2)$

(A) −41.76 (B) 41.76 (C) −39.26 (D) 1.4

Find the product of the below.

11. (-0.5)(0.53)

(A) -2.016 (B) -0.265 (C) -8.065 (D) 2.016

12. (10)(-2.51)

(A) -25.5 (B) -25.1 (C) 7.49 (D) 25.1

13. (-0.5)(1.9)

(A) 2.78 (B) 0.95 (C) -23.03 (D) -0.95

14. (-4)(-1.3)

(A) 5.2 (B) -4.7 (C) 7.6 (D) -5.3

15. (-3.6)(-5.3)

(A) -19.08 (B) 23.08 (C) 19.08 (D) -36.86

Simplify the below.

1. $4.7 \div 0.5$

(A) 5.2 (B) 9.4 (C) –2.35 (D) 2.35

2. $2.7 \div -5.4$

(A) –3.8 (B) 8.1 (C) –5.8 (D) –0.5

3. $2 \div -0.5$

(A) –6.9 (B) 2.5 (C) –4 (D) 0.25

4. $2.1 \div 0.3$

(A) 7 (B) –0.6 (C) –7 (D) –8.9

5. $-6.574 \div -0.1$

(A) 1.9 (B) –65.74 (C) 65.74 (D) –6.474

Simplify the below.

6. $8.8 \div 1.6$

(A) -5.5 (B) 7.2 (C) 5.5 (D) 10

7. $-0.4 \div 0.4$

(A) 4.97 (B) -1 (C) -9.5 (D) -6.4

8. $3.6 \div -9$

(A) -0.4 (B) -7.87 (C) -5.4 (D) -32.4

9. $4.1 \div -4.1$

(A) -16.81 (B) -1 (C) 0 (D) 16.81

10. $-7.8 \div 7.5$

(A) -6 (B) 3.1 (C) -1.04 (D) 2.6

Simplify the below.

11. $-9.2 \div -9.2$

(A) –84.64 (B) 1 (C) –1 (D) 1.3

12. $-1.7 \div -2$

(A) –2.1 (B) 6.2 (C) 2.5 (D) 0.85

13. $-7.7 \div 2.8$

(A) 21.56 (B) 2.91 (C) –2.75 (D) –4.9

14. $3 \div 0.6$

(A) –0.2 (B) 5 (C) –1.8 (D) –5.6

15. $1.2 \div 0.6$

(A) 2 (B) –1.6 (C) 1.8 (D) 0.72

Find the product of the below.

1. (6)(−7)(2)

(A) −85 (B) −84 (C) 84 (D) −90

2. (4)(−12)(11)

(A) −528 (B) −521 (C) −532 (D) −535

3. (−10)(−6)(−7)

(A) −415 (B) −420 (C) −412 (D) 420

4. (−14)(8)(4)

(A) −450 (B) −456 (C) −448 (D) −447

5. (−10)(8)(2)

(A) 0 (B) −167 (C) −160 (D) −173

Find the product of the below.

6. (0)(-3)(-8)

(A) 0 (B) -11 (C) 392 (D) -392

7. (6)(-4)(-7)

(A) -168 (B) -5 (C) 168 (D) 178

8. (-6)(9)(5)

(A) -260 (B) -280 (C) -270 (D) -968

9. (4)(-6)(-14)

(A) 332 (B) -16 (C) -336 (D) 336

10. (2)(3)(-8)

(A) -40 (B) -44 (C) -48 (D) 48

Find the product of the below.

11. (-3)(2)(0)

(A) 7 (B) -1 (C) 3 (D) 0

12. (-9)(-2)(13)

(A) -234 (B) 243 (C) 232 (D) 234

13. (-2)(5)(-1)

(A) 13 (B) -2 (C) 23 (D) 10

14. (11)(6)(-2)

(A) -132 (B) -142 (C) 132 (D) 15

15. (2)(-11)(-4)

(A) -88 (B) 82 (C) 77 (D) 88

Express the below in Scientific Notation.

1. $(9.1 \times 10^{0})(8 \times 10^{0})$

(A) 1.138×10^{-1}

(B) 7.28×10^{1}

(C) 1.138×10^{0}

(D) 7.28×10^{2}

2. $(9 \times 10^{5})(3.2 \times 10^{-1})$

(A) 2.88×10^{6}

(B) 28.8×10^{-2}

(C) 2.88×10^{5}

(D) 28.8×10^{6}

3. $(5 \times 10^{5})(3.1 \times 10^{4})$

(A) 15.5×10^{-11}

(B) 1.55×10^{10}

(C) 15.5×10^{10}

(D) 15.5×10^{11}

Express the below in Scientific Notation.

4. $(2.3 \times 10^{-4})(1.2 \times 10^{-3})$

(A) 1.917×10^{1}

(B) 2.76×10^{-7}

(C) 1.917×10^{-1}

(D) 2.76×10^{-8}

5. $(6.1 \times 10^{5})(5.1 \times 10^{-3})$

(A) 31.11×10^{3}

(B) 0.3111×10^{3}

(C) 1.196×10^{8}

(D) 3.111×10^{3}

6. $(5 \times 10^{-4})(2.7 \times 10^{-5})$

(A) 135×10^{-8}

(B) 1.35×10^{-8}

(C) 13.5×10^{-8}

(D) 1.852×10^{1}

Express the below in Scientific Notation.

7. $(6.9 \times 10^{-5})(9.61 \times 10^{-1})$

(A) 6.631×10^{4}

(B) 6.631×10^{-5}

(C) 7.18×10^{-5}

(D) 7.18×10^{-2}

8. $(8.8 \times 10^{-4})(8 \times 10^{-3})$

(A) 1.1×10^{0}

(B) 7.04×10^{-6}

(C) 1.1×10^{-1}

(D) 70.4×10^{-6}

9. $(9.8 \times 10^{-5})(4.4 \times 10^{5})$

(A) 4.312×10^{1}

(B) 2.227×10^{-10}

(C) 4.312×10^{2}

(D) 2.227×10^{2}

Express the below in Scientific Notation.

10. $(4.3 \times 10^{-3})(6.3 \times 10^{1})$

(A) 2.709×10^{2}

(B) 2.709×10^{1}

(C) 2.709×10^{-2}

(D) 2.709×10^{-1}

11. $(1.85 \times 10^{-1})(5.19 \times 10^{-3})$

(A) 9.602×10^{-4}

(B) 3.565×10^{1}

(C) 9.602×10^{-1}

(D) 9.602×10^{4}

12. $(7.1 \times 10^{-6})(6.6 \times 10^{5})$

(A) 0.4686×10^{-1}

(B) 0.4686×10^{0}

(C) 4.686×10^{0}

(D) 1.076×10^{-11}

Express the below in Scientific Notation.

13. $(2 \times 10^{2})(4 \times 10^{-3})$

(A) 5×10^{3}

(B) 8×10^{-1}

(C) 80×10^{-1}

(D) 5×10^{4}

14. $(7 \times 10^{3})(2.4 \times 10^{-5})$

(A) 2.917×10^{8}

(B) 0.168×10^{-1}

(C) 1.68×10^{-1}

(D) 0.0168×10^{-1}

15. $(2.2 \times 10^{-3})(6.4 \times 10^{-1})$

(A) 3.438×10^{-3}

(B) 1.408×10^{0}

(C) 1.408×10^{-3}

(D) 0.1408×10^{0}

Find the Greatest Common Factor (GCF) of the below.

1. 77, 88

(A) 33 (B) 5 (C) 11 (D) 616

2. 88, 55

(A) 440 (B) 11 (C) 22 (D) 5

3. 80, 48

(A) 16 (B) 240 (C) 2 (D) 80

4. 40, 92

(A) 20 (B) 920 (C) 3 (D) 4

5. 21, 49

(A) 147 (B) 35 (C) 7 (D) 28

Find the Greatest Common Factor (GCF) of the below.

6. 96, 32

(A) 2 (B) 32 (C) 8 (D) 96

7. 99, 77

(A) 6 (B) 11 (C) 3 (D) 693

8. 36, 24

(A) 4 (B) 12 (C) 72 (D) 2

9. 64, 80

(A) 16 (B) 8 (C) 4 (D) 320

10. 55, 75

(A) 4 (B) 5 (C) 25 (D) 825

Find the Greatest Common Factor (GCF) of the below.

11. 18, 63

(A) 126 (B) 3 (C) 9 (D) 18

12. 22, 33

(A) 2 (B) 66 (C) 9 (D) 11

13. 42, 28

(A) 12 (B) 84 (C) 14 (D) 7

14. 63, 84

(A) 7 (B) 21 (C) 3 (D) 252

15. 80, 28

(A) 2 (B) 4 (C) 560 (D) 8

Find the Greatest Common Factor (GCF) of the below.

1. $75x, 75y^2x, 30y^2x$

(A) $15xy$ (B) $3x$ (C) $15x$ (D) $150y^2x$

2. $40x^4, 64y^2x^2, 48x^4$

(A) $8x^2$ (B) $40x^2$ (C) $960x^4y^2$ (D) $8x$

3. $40y, 60y, 60xy$

(A) $20y$ (B) $120xy$ (C) $60y$ (D) $80y$

4. $16uv, 64v^2u, 44uv$

(A) $704v^2u$ (B) $4uv^2$ (C) $2uv$ (D) $4uv$

5. $23y^2, 10x, 52$

(A) 5 (B) 1 (C) 2 (D) $5980y^2x$

Find the Greatest Common Factor (GCF) of the below.

6. $36m^4, 72m^3, 54m^2$

(A) $2m^2$ (B) $216m^4$ (C) $8m^2$ (D) $18m^2$

7. $80n^2, 12mn^2, 16n$

(A) $2n$ (B) $4n^2$ (C) $240mn^2$ (D) $4n$

8. $36a, 27a^2b, 72a^3$

(A) $4a$ (B) $9a^2$ (C) $216a^3b$ (D) $9a$

9. $165v, 65u, 52v$

(A) 1 (B) 3 (C) $13u$ (D) $260vu$

10. $14, 42x, 56$

(A) $14y$ (B) 14 (C) 3 (D) $168x$

Find the Greatest Common Factor (GCF) of the below.

11. $80x^3, 60x^4, 40yx^2$

(A) $20x^2$ (B) $240yx^4$ (C) $19x^2$ (D) $4x^2$

12. $7y, 30, 26y$

(A) $2730y$ (B) x (C) 1 (D) y

13. $64v^2, 32v, 80u^2$

(A) $4v$ (B) $9u^2$ (C) 16 (D) $9uv$

14. $58ba^2, 26b^2a^2, 44ba^2$

(A) $2b^2a^2$ (B) $16588b^2a^2$ (C) $2ba^2$ (D) $2a^2$

15. $72ab^2, 45ab^2, 63ab^2$

(A) $2ab^2$ (B) $9a^2b^2$ (C) $2520ab^2$ (D) $9ab^2$

Find the Least Common Multiple (LCM) of the below.

1. | 32, 80 |

(A) 2560　　(B) 16　　(C) 40　　(D) 160

2. | 98, 91 |

(A) 1274　　(B) 7　　(C) 98　　(D) 8918

3. | 66, 44 |

(A) 2904　　(B) 660　　(C) 22　　(D) 132

4. | 55, 44 |

(A) 220　　(B) 10　　(C) 11　　(D) 2420

5. | 45, 100 |

(A) 4500　　(B) 900　　(C) 5　　(D) 50

Find the Least Common Multiple (LCM) of the below.

6. 51, 85

(A) 15 (B) 4335 (C) 255 (D) 17

7. 42, 56

(A) 2352 (B) 14 (C) 4 (D) 168

8. 27, 93

(A) 3 (B) 837 (C) 27 (D) 2511

9. 63, 79

(A) 1 (B) 2369 (C) 4977 (D) 9

10. 92, 16

(A) 1472 (B) 94 (C) 4 (D) 368

Find the Least Common Multiple (LCM) of the below.

11. | 91, 63 |

(A) 5733 (B) 13 (C) 7 (D) 819

12. | 52, 65 |

(A) 260 (B) 13 (C) 3380 (D) 26

13. | 80, 100 |

(A) 8000 (B) 20 (C) 2000 (D) 400

14. | 72, 54 |

(A) 3888 (B) 97 (C) 216 (D) 18

15. | 8, 92 |

(A) 4 (B) 184 (C) 736 (D) 121

Find the Least Common Multiple (LCM) of the below.

1. $75x^2, 96x^2, 84y$

(A) 12 (B) $1008yx^2$ (C) $16{,}800yx^2$ (D) $580608x^4y$

2. $52y^2x^2, 80x^3, 64x^2$

(A) $32y^2x^3$ (B) $4x^2$ (C) $4160y^2x^3$ (D) $266240y^2x^7$

3. $69a, 92a^2, 46b^2$

(A) 23 (B) $276b^2a^2$ (C) $292008a^3b^2$ (D) $276ba^2$

4. $60uv, 100, 80$

(A) 20 (B) $480000uv$ (C) $1200uv$ (D) $120uv$

5. $12ab^3, 68b^3, 52a^2b^2$

(A) $4b^2$ (B) $292a^2b^3$ (C) $2652a^2b^3$ (D) $42432a^3b^8$

Find the Least Common Multiple (LCM) of the below.

6. $21u^3, 63u^2v, 70u^2v$

(A) $630u^3v$ (B) $3150u^3v$ (C) $92610u^7v^2$ (D) $7u^2$

7. $8, 61v, 43$

(A) 1 (B) $20984v$ (C) $10492v$ (D) $41968v$

8. $96yx^2, 48x^2, 48y^2x^2$

(A) $96y^2x^2$ (B) $48x^2$ (C) $12y^2x^2$ (D) $221184y^3x^6$

9. $24vu^2, 66vu^2, 90u^3v$

(A) $3960u^3v^2$ (B) $6vu^2$ (C) $3960u^3v$ (D) $142560v^3u^7$

10. $42a^2, 63ba, 84a$

(A) $222264a^4b$ (B) $252a^3b$ (C) $252a^2b$ (D) $21a$

Find the Least Common Multiple (LCM) of the below.

11. $36uv, 72v, 60u^2v$

(A) $360u^2v$ (B) $1800u^2v$ (C) $12v$ (D) $155520u^3v^3$

12. $96a, 80, 64b^2$

(A) 16 (B) $491520ab^2$ (C) $960b^2a$ (D) $960b^3a$

13. $51y, 68y^2, 51x^2y$

(A) $4x$ (B) $9x^2$ (C) $216x^3y$ (D) $204x^2y^2$

14. $92ab, 92b, 46b^3$

(A) $4b^3a$ (B) $92b^3a$ (C) $46b$ (D) $389344ab^5$

15. $96x^2y, 80y^3, 58y$

(A) $445440x^2y^5$ (B) $13920x^2y^3$ (C) $2y$ (D) $13920x^2y^2$

Evaluate.

1. $((-20) \div (10 + 1 - 9)) \times 5$

(A) -50 (B) -56 (C) -49 (D) -40

2. $9^2 - (7 - 2 - 4)$

(A) 80 (B) 77 (C) 84 (D) 89

3. $(-1) + 8 + ((-6) - 7) \times (-2)$

(A) 34 (B) 32 (C) 39 (D) 33

4. $(21 - 9) \div ((-3) + 9) \times (-4)$

(A) -4 (B) 0 (C) -12 (D) -8

5. $9((-5) - 5) - (7 - (-3))$

(A) -103 (B) -96 (C) -91 (D) -100

Evaluate.

6. $9 \div (1 + 2) + 6 \div 2$

(A) 3 (B) 6 (C) 13 (D) -2

7. $((-21) \times 2) \div (3 - (-4)) \times (-5)$

(A) 21 (B) 30 (C) 27 (D) 39

8. $(9 + 9 + 10)(7 - 6)$

(A) 22 (B) 28 (C) 18 (D) 26

9. $(20 \times 2) \div (-10) - 3 - (-6)$

(A) -2 (B) 2 (C) 6 (D) -1

10. $7 - (-3) - ((-2) \div 2)^2$

(A) 11 (B) 12 (C) 9 (D) 16

Evaluate.

11. $10^2 + 2 - 8 \times 8$

(A) 28 (B) 38 (C) 32 (D) 48

12. $((-1) -1)(3 + 5^2)$

(A) -56 (B) -46 (C) -65 (D) -51

13. $10((-6) + 5 \times 4 - 4)$

(A) 99 (B) 105 (C) 100 (D) 102

14. $(-4) \div ((-2)(9 - (-1) - 9))$

(A) 2 (B) 8 (C) -3 (D) -1

15. $((-4) - (-3)) \times 12 \div (1 - 3)$

(A) 0 (B) 6 (C) 13 (D) 1

Express the below statement as an algebraic expression.

1. The quotient of a number and 8 is equal to 9

(A) $8^2 \geq 9$ (B) $\frac{8}{n} = 9$ (C) $\frac{n}{8} = 9$ (D) $8 + n > 9$

2. The quotient of y and 6 is less than or equal to 17

(A) $\frac{6}{y} \leq 17$ (B) $\frac{y}{6} \leq 17$ (C) $\frac{6}{y} = 17$ (D) $6 + y \leq 17$

3. The sum of a number and 7 is 36

(A) $n^7 = 36$ (B) $7 - n = 36$ (C) $n + 7 = 36$ (D) $n^3 = 36$

4. The sum of a number and 9 is equal to 43

(A) $\frac{9}{2} = 43$ (B) $2n < 43$ (C) $n \cdot 9 > 43$ (D) $n + 9 = 43$

5. n decreased by 18 is less than or equal to 14

(A) $18n \leq 14$ (B) $n - 18 \leq 14$ (C) $18^3 \leq 14$ (D) $18 - n$

Express the below statement as an algebraic expression.

6. 11 more than a is greater than or equal to 20

(A) $11a \geq 20$ (B) $a + 11 \geq 20$ (C) $a^{11} \geq 20$ (D) $11 - a \leq 20$

7. 3 less than a is greater than or equal to 42

(A) $a \cdot 3 \geq 42$ (B) $\frac{a}{2} \geq 42$ (C) $a - 3 \geq 42$ (D) $3^3 \geq 42$

8. The quotient of x and 6 is 33

(A) $6 + x = 33$ (B) $\frac{x}{6} = 33$ (C) $x - 6 = 33$ (D) $\frac{6}{x} = 33$

9. The sum of a and 12 is 41

(A) $a^{12} = 41$ (B) $12a = 41$ (C) $12^2 = 41$ (D) $a + 12 = 41$

10. The product of x and 5 is equal to 28

(A) $x \cdot 5 = 28$ (B) $x^5 = 2$ (C) $x + 5 = 28$ (D) $5 - x \leq 28$

Express the below statement as an algebraic expression.

11. The sum of u and 5 is 6

(A) $u - 5 = 6$ (B) $5^2 \geq 6$ (C) $u + 5 = 6$ (D) $2u > 6$

12. t decreased by 26 is less than 35

(A) $t^{26} < 35$ (B) $t - 26 < 35$ (C) $26 - t < 35$ (D) $26^t < 35$

13. The product of a number and 10 is 17

(A) $n + 10 = 17$ (B) $n \cdot 10 = 17$ (C) $n^{10} = 17$ (D) $\frac{n}{10} = 17$

14. The product of a number and 7 is less than 8

(A) $n \cdot 7 < 8$ (B) $n^7 > 8$ (C) $n^2 < 8$ (D) $7 + n < 8$

15. The quotient of a number and 2 is less than 20

(A) $\frac{2}{n} < 20$ (B) $2 + n < 20$ (C) $\frac{n}{2} < 20$ (D) $n \cdot 2 \geq 20$

Write the verbal expression of the inequation.

1. $n \cdot 8 < 18$

(A) The product of a number and 8 is less than 18.
(B) 8 plus a number is less than 18.
(C) 8 decreased by a number is less than 18.
(D) Half of 8 is less than 18.

2. $n \cdot 5 = 24$

(A) A number squared is 24.
(B) The product of a number and 5 is 24.
(C) Half of a number is 24.
(D) A number decreased by 5 is 24.

3. $n + 7 = 7$

(A) 7 less than a number is 7.
(B) 7 to the n is 7.
(C) The product of a number and 7 is 7.
(D) The sum of a number and 7 is 7.

Write the verbal expression of the inequation.

4. $n \cdot 7 = 34$

(A) A number times 7 is equal to 34.
(B) A number cubed is equal to 34.
(C) The 7th power of a number is equal to 34.
(D) 7 squared is equal to 34.

5. $\frac{n}{5}$

(A) 5 times a number.
(B) 5 divided by a number.
(C) A number divided by 5.
(D) 5 increased by a number.

6. $n - 5 \geq 7$

(A) A number cubed is greater than or equal to 7.
(B) Half of a number is greater than or equal to 7.
(C) 5 increased by a number is greater than or equal to 7.
(D) The difference of a number and 5 is greater than or equal to 7.

Write the verbal expression of the inequation.

7. $\frac{72}{n}$

(A) The quotient of 72 and a number.
(B) 72 times a number.
(C) 72 increased by a number.
(D) The quotient of a number and 72.

8. $n + 10 < 33$

(A) The product of 10 and a number is less than 33.
(B) A number increased by 10 is less than 33.
(C) A number decreased by 10 is less than 33.
(D) 10 minus a number is less than 33.

9. $\frac{40}{5}$

(A) 5 increased by 40.
(B) 40 minus 5.
(C) 5 divided by 40.
(D) 40 divided by 5.

Write the verbal expression of the inequation.

10. $16 - 11$

(A) 16 times 11.
(B) 11 less than 16.
(C) 16 divided by 11.
(D) 11 cubed.

11. $\frac{n}{2} = 24$

(A) A number squared is equal to 24.
(B) A number decreased by 2 is equal to 24.
(C) 2 to the n is equal to 24.
(D) Half of a number is equal to 24.

12. $n + 6 = 37$

(A) The n power of 6 is equal to 37.
(B) Twice 6 is equal to 37.
(C) A number plus 6 is equal to 37.
(D) 6 times a number is equal to 37.

Write the verbal expression of the inequation.

13. $n + 8 = 47$

(A) The n power of 8 is equal to 47.
(B) A number increased by 8 is equal to 47.
(C) The product of a number and 8 is equal to 47.
(D) 8 decreased by a number is equal to 47.

14. $n - 6 \geq 41$

(A) 6 minus a number is greater than or equal to 41.
(B) Twice 6 is greater than or equal to 41.
(C) A number minus 6 is greater than or equal to 41.
(D) 6 times a number is greater than or equal to 41.

15. $n \cdot 5 \leq 45$

(A) The product of a number and 5 is less than or equal to 45.
(B) 5 plus a number is less than or equal to 45.
(C) 5 minus a number is less than or equal to 45.
(D) A number more than 5 is less than or equal to 45.

Simplify the below expression.

1. $-4(4 + 8x) + 8x$

(A) $-16 - 23x$ (B) $-16 - 24x$ (C) $-5 + 32x$ (D) $-6 + 32x$

2. $8 - (5n + 8)$

(A) $10n - 16$ (B) $10n - 12$ (C) $-5n$ (D) $10n - 11$

3. $-(6 - 5p) - 4$

(A) $-10 + 5p$ (B) $-15 - 3p$ (C) $-10 - 3p$ (D) $-7 - 3p$

4. $-2n - 5(n - 1)$

(A) $-23n - 35$ (B) $-7n + 3$ (C) $-7n + 5$ (D) $-29n - 35$

5. $-3 - 8(7p - 6)$

(A) $4 - 5p$ (B) $-2 - 5p$ (C) $45 - 56p$ (D) $25 - 16p$

Simplify the below expression.

6. $-7(x - 2) - 6$

(A) $-6x + 26$ (B) $-10x - 15$ (C) $2x + 24$ (D) $-7x + 8$

7. $7x - 7(5x - 4)$

(A) $-28x + 24$ (B) $7 + 8x$ (C) $-28x + 28$ (D) $3 + 8x$

8. $-2(a + 3) + 1$

(A) $-15 - 5a$ (B) $-11 - 5a$ (C) $-2a - 5$ (D) $-18 - 5a$

9. $-2x - 6(x + 3)$

(A) $4x - 18$ (B) $-8x - 18$ (C) $-x - 18$ (D) $4x - 10$

10. $-8(-5 - 7k) + 4$

(A) $2 - 8k$ (B) $44 + 56k$ (C) $-4 - 38k$ (D) $-28 - 18k$

Simplify the below expression.

11. $-6 - 3(1 - 5v)$

(A) $-9 + 15v$ (B) $-8 + 15v$ (C) $-4v - 13$ (D) $-1 + 15v$

12. $-6(8 - v) - 7v$

(A) $7v - 48$ (B) $-48 - v$ (C) $7v - 40$ (D) $-48 - 8v$

13. $-7(x + 6) - 2$

(A) $-7x - 44$ (B) $-20 + 5x$ (C) $-21 + 5x$ (D) $-16x - 2$

14. $3p - 7(1 + p)$

(A) $-4p - 7$ (B) $-14p + 12$ (C) $11 - 56p$ (D) $24 - 15p$

15. $-6 - 8(1 + 2n)$

(A) $19n - 64$ (B) $10n + 14$ (C) $19n - 56$ (D) $-14 - 16n$

Solve the below inequality.

1. $-7x - 5(x - 1) \le 89$

(A) No solution

(B) $x \ge -10$

(C) $x \ge -7$

(D) $x \ge -6$

2. $123 > -3 + 6(1 - 5x)$

(A) $x > -4$

(B) $x > -23$

(C) $x > -34$

(D) $x > -5$

3. $5(5x - 3) + 8 < -132$

(A) $x < -39$

(B) $x < -10$

(C) $x < -5$

(D) $x < -35$

4. $282 < -6 - 8(-1 + 5k)$

(A) No solution

(B) {All real numbers}

(C) $k < -37$

(D) $k < -7$

Solve the below inequality.

5. $7(x - 7) \leq -84$

(A) $x \leq -12$

(B) $x \leq -5$

(C) $x \leq -40$

(D) $x \leq -28$

6. $87 \leq -4(-6 + 2r) - r$

(A) $r \geq -2$

(B) $r \leq -7$

(C) $r \leq -16$

(D) $r \geq -16$

7. $-84 \geq 7(-6 + 2k)$

(A) $k \leq 2$

(B) $k \leq -3$

(C) $k \geq -3$

(D) $k \leq -27$

8. $-105 < -5(1 - 5p)$

(A) $p > -8$

(B) $p > -22$

(C) $p > -24$

(D) $p > -4$

Solve the below inequality.

9. $-98 \geq -7(-2 + 2x)$

(A) $x \geq -27$

(B) $x \geq -8$

(C) $x \geq -39$

(D) $x \geq 8$

10. $396 \geq -6v + 6(2 - 7v)$

(A) $v \geq 1$

(B) $v \geq -8$

(C) $v \leq -26$

(D) $v \leq 1$

11. $3(-7a - 7) \leq -84$

(A) $a \geq -22$

(B) {All real numbers.}

(C) $a \leq 3$

(D) $a \geq 3$

12. $-7 - 3(1 + 4r) \leq -106$

(A) $r \geq 8$

(B) $r \geq 0$

(C) $r \geq -13$

(D) $r \geq -10$

Solve the below inequality.

13. $2(-1 - 8k) < -82$

(A) $k < -14$

(B) $k > -1$

(C) $k > -14$

(D) $k > 5$

14. $4(6 - 8r) \geq 248$

(A) $r \leq -22$

(B) $r \geq -1$

(C) $r \leq -7$

(D) $r \geq -22$

15. $-82 < 8(r - 5) - 2r$

(A) $r < -26$

(B) $r < -7$

(C) $r > -7$

(D) $r < -30$

Evaluate the below expression.

1. $m(p + (-2)^3 + q)$; use $m = -4$, $p = -10$, and $q = 2$

(A) 69 (B) 63 (C) 64 (D) 56

2. $p(1 - m) - \frac{m}{4}$; use $m = 4$, and $p = 7$

(A) −22 (B) −31 (C) −17 (D) −27

3. $p(q - (-2q - p))$; use $p = -4$, and $q = 8$

(A) −80 (B) −71 (C) −79 (D) −70

4. $y(y + y + z + 3)$; use $y = 7$, and $z = -8$

(A) 55 (B) 63 (C) 70 (D) 69

5. $x(2 - y + |y|)$; use $x = 9$, and $y = 9$

(A) 18 (B) 16 (C) 25 (D) 11

Evaluate the below expression.

6. $h - h + 4 + j^2$; use $h = -3$, and $j = 7$

(A) 49 (B) 53 (C) 59 (D) 46

7. $|p + m|\,(p - 5)$; use $m = 5$, and $p = -1$

(A) -19 (B) -24 (C) -14 (D) -25

8. $zx\,\dfrac{y - y}{6} + 5$; use $y = -9$, and $z = -10$

(A) 3 (B) 6 (C) 8 (D) 5

9. $\left(\dfrac{h}{6}\right)^2 + j - j$; use $h = -6$, and $j = 7$

(A) 9 (B) 1 (C) 7 (D) -5

10. $x + y + x - |y|$; use $x = -10$, and $y = -5$

(A) -30 (B) -28 (C) -33 (D) -24

Evaluate the below expression.

11. $y + x^2(x + 4)$; use $x = -2$, and $y = -4$

(A) −3 (B) 11 (C) 1 (D) 4

12. $\frac{(y - y)^2 + x}{2}$; use $x = 10$, and $y = 8$

(A) −10 (B) −4 (C) 11 (D) 5

13. $p - (m - 2 + m - m))$; use $m = 1$, and $p = 3$

(A) 4 (B) 8 (C) 13 (D) 7

14. $m - \frac{mp}{2} + p$; use $m = 5$, and $p = 6$

(A) 6 (B) 5 (C) 12 (D) -4

15. $-\frac{10}{2}(|y| + x)$; use $x = -4$, and $y = 9$

(A) −21 (B) −20 (C) −31 (D) −25

Evaluate the below expression.

1. $\frac{n}{77} = \frac{37}{77}$

(A) -39 (B) 37 (C) 4 (D) 66

2. $488 = -61r$

(A) -6 (B) 55 (C) 90 (D) -8

3. $40n = 2240$

(A) -44 (B) -22 (C) 56 (D) -14

4. $-46m = 3680$

(A) -99 (B) -80 (C) 9 (D) 97

5. $\frac{13}{19} = \frac{x}{76}$

(A) 97 (B) 52 (C) 44 (D) 23

Evaluate the below expression.

6. $-1892 = 44x$

(A) -87 (B) 63 (C) -43 (D) -86

7. $7857 = -81k$

(A) -97 (B) -85 (C) -53 (D) 79

8. $2592 = 72x$

(A) 58 (B) -76 (C) 36 (D) 13

9. $74 = x + 48$

(A) 43 (B) 26 (C) 11 (D) $1\frac{13}{24}$

10. $-4v = -244$

(A) {-3} (B) {36} (C) {61} (D) {-92}

Evaluate the below expression.

11. $\frac{m}{100} = 86$

(A) $\frac{43}{50}$ (B) 8600 (C) 41 (D) -14

12. $-81 = -59 - b$

(A) $1\frac{22}{59}$ (B) 22 (C) -22 (D) -99

13. $p + (-54) = -147$

(A) $2\frac{13}{18}$ (B) 12 (C) -93 (D) -6

14. $-10 - x = 77$

(A) -87 (B) $-7\frac{7}{10}$ (C) 87 (D) 67

15. $-41 = \frac{x}{80}$

(A) -3280 (B) -35 (C) 39 (D) $-\frac{41}{80}$

Solve the below equation.

1. $-176 = 8(4k - 2)$

(A) {-16} (B) {-8}

(C) {-5} (D) {-1}

2. $-3(5x - 6) = 93$

(A) {13} (B) {-14}

(C) No solution (D) {-5}

3. $-84 = 7(n - 5)$

(A) {8} (B) {5}

(C) {-7} (D) {16}

4. $-4(1 - 4n) = -116$

(A) {-16} (B) {-7}

(C) {-1} (D) {-12}

Solve the below equation.

5. $97 = -6x + 7(-6x + 7)$

(A) {6}

(B) {All real numbers.}

(C) {-1}

(D) {16}

6. $5(1 - 6a) = 155$

(A) {-2}

(B) {-5}

(C) {7}

(D) {3}

7. $-136 = -8(3a + 2)$

(A) {-6}

(B) {5}

(C) {4}

(D) {9}

8 $-4(-7k - 8) = 116$

(A) {All real numbers.}

(B) {-3}

(C) {3}

(D) {12}

Solve the below equation.

9. $-4(5p + 8) = 128$

(A) {14}

(B) {-8}

(C) {-12}

(D) {16}

10. $3(1 - 4x) = 99$

(A) No solution

(B) {-8}

(C) {9}

(D) {All real numbers.}

11. $-3(3x + 5) = -87$

(A) {-12}

(B) {-7}

(C) {8}

(D) {-3}

12. $8(6a + 2) = -128$

(A) {-5}

(B) {1}

(C) {-3}

(D) {-7}

Solve the below equation.

13. $329 = -7(1 + 8a)$

(A) {-13}
(B) {-9}
(C) {-6}
(D) {4}

14. $-8(7x - 3) - 8x = 344$

(A) {-5}
(B) {11}
(C) {0}
(D) {All real numbers.}

15. $168 = 7(3x + 3)$

(A) {-6}
(B) {All real numbers.}
(C) {7}
(D) {16}

Evaluate the below expressions using the values given.

1. $7 + p - m$; use $m = -8$, $p = 7$

(A) 25 (B) 26 (C) 22 (D) 20

2. $c + a - c$; use $a = -4$, and $c = -10$

(A) −4 (B) −2 (C) −13 (D) 4

3. $|n| + m$; use $m = 9$, and $n = -1$

(A) 8 (B) 16 (C) 10 (D) 17

4. $\frac{z}{4} + x$; use $x = -8$, and $z = -8$

(A) −10 (B) 0 (C) −6 (D) −2

5. $y(z + y)$; use $y = 2$, and $z = 6$

(A) 24 (B) 16 (C) 12 (D) 15

Evaluate the below expressions using the values given.

6. xy^3; use x = −1, and y = 4

(A) −64 (B) −70 (C) −66 (D) −72

7. $(x + y)^2$; use x = 7, and y = −3

(A) 26 (B) 22 (C) 6 (D) 16

8. −3m + n; use m = −3, and n = 10

(A) 19 (B) 24 (C) 26 (D) 14

9. $p^2 - m$; use m = 1, and p = −2

(A) 3 (B) 13 (C) -3 (D) -6

10. p + 10q; use p = −9, and q = −4

(A) −42 (B) −49 (C) −56 (D) −53

Evaluate the below expressions using the values given.

11. $j + |k|$; use $j = -4$, and $k = 5$

(A) –9 (B) 1 (C) –20 (D) –1

12. $p + p + q$; use $p = -9$, and $q = 6$

(A) –3 (B) –6 (C) –14 (D) –12

13. $x - 6 + y$; use $x = -3$, and $y = -10$

(A) –19 (B) –12 (C) –29 (D) –26

14. $h + j - 9$; use $h = -9$, and $j = 4$

(A) –8 (B) –11 (C) –15 (D) –14

15. $|m| - p$; use $m = -2$, and $p = -6$

(A) 3 (B) 8 (C) 12 (D) –2

Solve the below equation.

1. $|7 - 7m| = 63$

(A) $\{-8, 10\}$

(B) $\left\{10, -\frac{74}{7}\right\}$

(C) $\left\{\frac{24}{7}, -4\right\}$

(D) $\{5, -10\}$

2. $|x - 6| = 5$

(A) $\{7, -1\}$

(B) $\{11, 1\}$

(C) $\left\{-2, \frac{1}{3}\right\}$

(D) $\left\{-8, \frac{49}{5}\right\}$

3. $|n + 6| = 1$

(A) $\left\{-3, \frac{9}{7}\right\}$

(B) $\{9, -11\}$

(C) $\{-5, -7\}$

(D) $\{-8, -2\}$

Solve the below equation.

4. $|-6 + 7m| = 50$

(A) $\left\{-3, \frac{5}{2}\right\}$

(B) $\{8\}$

(C) $\left\{3, \frac{1}{3}\right\}$

(D) $\left\{8, -\frac{44}{7}\right\}$

5. $|6m - 7| = 43$

(A) $\left\{\frac{25}{3}, -6\right\}$

(B) $\left\{\frac{47}{5}, -9\right\}$

(C) $\left\{7, -\frac{13}{2}\right\}$

(D) $\left\{\frac{25}{3}\right\}$

6. $|-7x - 9| = 16$

(A) $\{-16, 8\}$

(B) $\{-16\}$

(C) $\left\{-\frac{1}{3}, 3\right\}$

(D) $\left\{-\frac{25}{7}, 1\right\}$

Solve the below equation.

7. $|9x - 4| = 4$

(A) $\left\{\frac{8}{9}, 0\right\}$

(B) $\left\{\frac{8}{9}\right\}$

(C) $\left\{-\frac{17}{2}, 8\right\}$

(D) $\left\{7, -\frac{5}{2}\right\}$

8. $|-9a - 2| = 2$

(A) $\{4, 3\}$

(B) $\left\{-\frac{4}{9}, 0\right\}$

(C) $\{4\}$

(D) $\{4, -6\}$

9. $|10 - 5b| = 20$

(A) $\{-2, 6\}$

(B) $\{-2\}$

(C) $\{-6\}$

(D) $\{-6, 5\}$

Solve the below equation.

10. $|2m - 1| = 13$

(A) $\{13, 7\}$

(B) $\{7, -6\}$

(C) $\{7, -11\}$

(D) $\left\{-1, \frac{11}{4}\right\}$

11. $|8x + 2| = 78$

(A) $\left\{\frac{19}{2}\right\}$

(B) $\{-3, -7\}$

(C) $\{4, -8\}$

(D) $\left\{\frac{19}{2}, -10\right\}$

12. $|3v + 8| = 2$

(A) $\left\{-\frac{39}{5}, 5\right\}$

(B) $\left\{-2, -\frac{10}{3}\right\}$

(C) $\left\{2, \frac{4}{3}\right\}$

(D) $\left\{4, -\frac{34}{7}\right\}$

Solve the below equation.

13. $|-6n - 4| = 28$

(A) $\left\{8, -\frac{42}{5}\right\}$

(B) $\left\{8, -\frac{48}{5}\right\}$

(C) $\left\{-\frac{16}{3}, 4\right\}$

(D) $\{8\}$

14. $|9x - 6| = 30$

(A) $\left\{\frac{9}{4}, -4\right\}$

(B) $\left\{4, -\frac{8}{3}\right\}$

(C) $\{4\}$

(D) $\left\{6, -\frac{9}{2}\right\}$

15. $|3n - 8| = 22$

(A) $\{-7, -5\}$

(B) $\left\{7, -\frac{37}{5}\right\}$

(C) $\left\{2, \frac{2}{7}\right\}$

(D) $\left\{10, -\frac{14}{3}\right\}$

Round your answer to the nearest whole number.

1. Sri is traveling to London. How many Euros will she receive if she exchanged $18 at the rate of 3 Euros to $1 ?

   (A) £6 (B) £54 (C) £4 (D) £2

2. Kate bought one bag of mangoes for $2. How many bags of mangoes can Kate buy for $6 ?

   (A) 12 (B) 2 (C) 3 (D) 1

3. Derick enlarged the size of a rectangle to a width of 4 in. What is the new height if it was originally 2 in wide and 4 in tall ?

   (A) 9 in (B) 7 in (C) 8 in (D) 10 in

4. The currency in Huwaei is the Hans. Gary is traveling from Florida to Huwaei. The Currency exchange rate is $1 to 6 Hans. At this rate, how many Hans will he get if he has exchanged $200 ?

   (A) 1300 Hans (B) 1400 Hans (C) 1200 Hans (D) 1000 Hans

Round your answer to the nearest whole number.

5. Maya is enlarging the size of her prototype to a height of 10 in. What is the new width if it was originally 2 in tall and 4 in wide ?

(A) 20 in (B) 1 in (C) 50 in (D) 2 in

6. A box of blueberries cost $4. How many boxes of blueberries can Mary but for $16 ?

(A) 5 (B) 2 (C) 4 (D) 64

7. A rose plant cost $2. How many rose plants can Nathan buy for $12 ?

(A) 9 (B) 6 (C) 7 (D) 24

8. A picture is 3 cm wide and 2 cm tall. If it is enlarged to a width of 18 cm, then what is the new height ?

(A) 12 cm (B) 108 cm (C) 3 cm (D) 11 cm

Round your answer to the nearest whole number.

9. A picture is 2 mm wide and 3 mm tall. If it is enlarged to a height of 6 mm, find the new width ?

(A) 4 mm (B) 12 mm (C) 1 mm (D) 3 mm

10. A box of Pears costs $3. How many box of Pears can Jack buy for $6 ?

(A) 2 (B) 18 (C) 3 (D) 1

11. Rosy reduced the size of a painting to a width of 3 cm. What is the new height if it was originally 3 cm tall and 9 cm wide ?

(A) 1 cm (B) 9 cm (C) 2 cm (D) 3 cm

12. A box of cherries costs $2. How many boxes can sam buy for $8 ?

(A) 3 (B) 4 (C) 16 (D) 1

Round your answer to the nearest whole number.

13. Darryl enlarged the size of a photo to a width of 12 in. What is the new height if it was originally 3 in wide and 2 in tall ?

(A) 8 in (B) 48 in (C) 7 in (D) 1 in

14. The money used in Alaysia is called the Ginggit. The exchange rate is 4 Ginggits for every $1. Find how many Ginggits you would receive if you exchanged $2 ?

(A) 8 Ginggits (B) 7 Ginggits (C) 1 Ginggits (D) 9 Ginggits

15. A box of Apples costs $2. How many boxes of Apples can Tim buy for $18 ?

(A) 2 (B) 4 (C) 9 (D) 3

Evaluate the below proportion.

1. $\frac{8}{4} = -\frac{4}{k}$

(A) {6.7} (B) {5.9} (C) {-2} (D) {-4.2}

2. $\frac{10}{9} = \frac{2}{b}$

(A) {1.8} (B) {3.6} (C) {4} (D) {-9.1}

3. $-\frac{3}{9} = \frac{n}{2}$

(A) {9.8} (B) {-5} (C) {-0.67} (D) {-9.1}

4. $\frac{5}{n} = -\frac{7}{10}$

(A) {-7.14} (B) {-9.1} (C) {-3.9} (D) {-7.3}

Evaluate the below proportion.

5. $\frac{7}{8} = -\frac{4}{b}$

(A) {4.7} (B) {5} (C) {-4.57} (D) {5.7}

6. $\frac{n}{9} = \frac{7}{8}$

(A) {5} (B) {-9.7} (C) {5.3} (D) {7.88}

7. $\frac{2}{5} = \frac{x}{9}$

(A) {3.6} (B) {-4} (C) {5.1} (D) {-4.2}

8. $\frac{7}{m} = \frac{9}{3}$

(A) {2.33} (B) {-8.532} (C) {-9} (D) {1.5}

Evaluate the below proportion.

9.

$$\frac{2}{8} = \frac{p}{3}$$

(A) {2} (B) {0.75} (C) {4.6} (D) {6.3}

10.

$$-\frac{n}{2} = -\frac{7}{6}$$

(A) {-8.6} (B) {-2} (C) {-5.2} (D) {2.33}

11.

$$\frac{8}{4} = \frac{x}{10}$$

(A) {1} (B) {-2} (C) {20} (D) {4.1}

12.

$$\frac{2}{x} = \frac{4}{2}$$

(A) {3.8} (B) {-5.1} (C) {1} (D) {7.7}

Evaluate the below proportion.

13. $$-\frac{9}{10} = -\frac{x}{3}$$

(A) {4.5} (B) {2.7} (C) {-3.9} (D) {-3.3}

14. $$\frac{6}{5} = \frac{7}{a}$$

(A) {2} (B) {5.83} (C) {4.3} (D) {8.93}

15. $$-\frac{6}{2} = \frac{10}{x}$$

(A) {7.7} (B) {10} (C) {-8.6} (D) {-3.33}

1. Olivia wants to buy a bag priced at \$52.50. Store is offering a discount of 30% today. Find the discount amount of the bag.

   (A) \$63.00 (B) \$3.75 (C) \$42.00 (D) \$15.75

2. Johnson bought a magazine at 10% discount which was priced at \$1 What is the selling price of the magazine ?

   (A) \$1.10 (B) \$0.90 (C) \$0.10 (D) \$1.20

3. Maria had a coupon of 50% discount. She selected a dress priced at \$15.95. What will be the selling price of the dress after discount.

   (A) \$23.92 (B) \$18.34 (C) \$7.90 (D) \$7.98

4. Monopoly game costs \$25.00. The store is offering a discount of 30% on it. What is the selling price of the game ?

   (A) \$26.25 (B) \$7.50 (C) \$32.50 (D) \$17.50

5. Mia wants to go to concert and the tickets are priced at $184.50. Her friend gave her a coupon of 35% on the concert tickets. How much does the ticket costs after the discount ?

(A) $119.93 (B) $166.05 (C) $119.90 (D) $64.58

6. A shop is offering 20% discount for one day on a video game that is priced at $9.99? What is the selling price of the video game after discount ?

(A) $8 (B) $7.5 (C) $6.95 (D) $7.09

7. Cathy purchased a bag priced at $37.95. She received a discount of 40% at the billing counter as a preferred customer. What is the selling price of the bag ?

(A) $53.13 (B) $22.77 (C) $36.05 (D) $15.18

8. The original price of a music CD is $22.50 and discount offered on it is 30%. What is the selling price of the music CD ?

(A) $6.75 (B) $21.38 (C) $15.75 (D) $29.25

9. A new X - Box game is costing $99.50. Sofia bought the X - Box game at the discounted price of 53%. How much did Sofia paid ?

(A) $94.52 (B) $152.24 (C) $52.73 (D) $46.77

10. A printing paper supply box costs $200 with a discount of 54% on it. What is the selling price of the printing paper box ?

(A) $92 (B) $0.95 (C) $95 (D) $89

11. A box of Apples costs $14.50. Food mart is offering 25% discount on it today. Find the selling price of the box of Apples.

(A) $10.88 (B) $13.77 (C) $15.95 (D) $3.62

12. A box of Oranges costs $14.50. Food mart is offering 7% discount on it today. Find the selling price of the box of Oranges?

(A) $16.45 (B) $13.09 (C) $13.49 (D) $12.05

13. Stella purchased a Jar for $0.95. What is the selling price if the discount offered on the jar is 50% ?

(A) $1.14 (B) $0.40 (C) $1.42 (D) $0.48

14. What is the selling price of a puzzle that is priced $40.00 and when a discount offered on it is 20% ?

(A) $32.00 (B) $48.00 (C) $44.00 (D) $8.00

15. A box of pencils costs $2.95. A discount of 50% is offered on it today. What is the selling price of the box of pencils.

(A) $3.39 (B) $1.48 (C) $4.43 (D) $2.80

1. Lydia purchased a Bag at $22.99 and sells it at a 40% price increase. Find the selling price of the Bag.

(A) $21.84 (B) $9.20 (C) $32.19 (D) $13.79

2. Robert purchased a X - Box at $629.99 and sells it at a 55% price increase. Find the selling price of the X - Box.

(A) $566.99 (B) $346.49 (C) $283.50 (D) $976.48

3. Cathy purchased a Play Station at $219.99 and sells it at a 20% price increase. Find the selling price of the Play Station.

(A) $175.99 (B) $252.99 (C) $230.99 (D) $263.99

4. Kim purchased a Monopoly game for $19.95 and sold at 90% price hike. What is the selling price of the monopoly game ?

(A) $17.95 (B) $37.91 (C) $37.99 (D) $1.99

5. Bryan purchased an Apartment at $54,000.00 and sells it at a 40% price hike. Find the selling price of the Apartment.

(A) $64,800.00 (B) $62,100.00 (C) $75,600.00 (D) $75,000.00

6. Roma purchased a Furniture set at $8,400.00 and sells it at a 80% price increase. What is the selling price of the Furniture set ?

(A) $15,120.00 (B) $7,980.00 (C) $6,720.00 (D) $1,680.00

7. Albert purchased a Scrabble game at $20.50 and sells it at a 20% price increase. Find the selling price of the Scrabble game.

(A) $4.10 (B) $24.60 (C) $17.43 (D) $16.40

8. Jack purchased a Nintendo game at $48.95 and sells it at a 55% price increase. Find the selling price of the Nintendo game.

(A) $41.61 (B) $75.87 (C) $22.03 (D) $26.92

9. Rachel purchased a Lego set at $99.95 and sells it at a 20% price increase. Find the selling price of the Lego set.

(A) $84.96 (B) $19.99 (C) $119.94 (D) $79.96

10. Alex purchased a Music CD at $14.99 and sells it at a 23% price increase. Find the selling price of the Music CD.

(A) $11.54 (B) $16.49 (C) $3.45 (D) $18.44

11. Andy purchased a Harry potter Book at $4.95 and sells it at a 10% price increase. Find the selling price of the Harry potter Book.

(A) $4.21 (B) $4.70 (C) $0.50 (D) $5.45

12. Anthony purchased a box of pens at $2.00 and sells it at a 75% price increase. Find the selling price of the box of pens.

(A) $1.90 (B) $0.50 (C) $1.50 (D) $3.50

13. Brad purchased a Office supplies at $160.00 and sells it at a 80% price increase. Find the selling price of the Office supplies.

(A) $288.00 (B) $32.00 (C) $128.00 (D) $152.00

14. Larry purchased a Pokemon Game at $23.00 and sells it at a 75% price increase. Find the selling price of the Pokemon Game.

(A) $19.55 (B) $40.25 (C) $17.25 (D) $5.75

15. Tony purchased a bag of lemons at $2.95 and sells it at a 20% price increase. Find the selling price of the bag of lemons.

(A) $0.59 (B) $3.54 (C) $2.66 (D) $2.36

1. Claire is purchasing a land priced at $16,500 and she pays a tax of 5%. Find the selling price of the land.

(A) $13,200.00 (B) $14,025.00 (C) $17,325.00 (D) $18,150.00

2. Richard purchased a box of strawberries for $9.50 and 6% tax is levied on that. What is the selling price of the box of strawberries ?

(A) $10.07 (B) $8.55 (C) $8.93 (D) $0.57

3. When 5% of tax is added to a milk can of $10.00. What is the sale price.

(A) $10.50 (B) $0.50 (C) $9.50 (D) $11.50

4. The original price of a car is $37,000.00 and the tax need to be paid for the bike is 6%, what is the selling price of the car ?

(A) $2,220.00 (B) $39,220.00 (C) $34,780.00 (D) $42,550.00

5. Maria purchased dining table set for \$64.95. Find out the selling price if the tax of 2% is added to it ?

(A) \$61.70 (B) \$63.65 (C) \$1.30 (D) \$66.25

6. Original price of a chocolate bar is \$2.20 and tax of 1% is added to the price. Find out the sale price of it ?

(A) \$2.22 (B) \$2.12 (C) \$2.02 (D) \$1.20

7. Jill selected a X - Box for his son that is tagged with price \$499.95. While paying the bill, he noticed that 6% tax is added to the tagged price. What could be the amount he paid in total for the X - Box ?

(A) \$520.95 (B) \$509.995 (C) \$529.947 (D) \$499.98

8. The original price of a dress is \$59.95. Calculate the selling price if 6% tax is added to it ?

(A) \$63.55 (B) \$56.95 (C) \$56.35 (D) \$65.95

9. Charlie purchased a bag for $34.95 and paid 6% tax for the price. Find out the selling price of the bag ?

(A) $29.71 (B) $32.85 (C) $37.05 (D) $2.10

10. When 4% of tax is levied on a shirt priced at $19.50, what could be the selling price of it ?

(A) $0.78 (B) $20.28 (C) $16.57 (D) $18.72

11. If $1.50 is the original price of a magazine, calculate the selling price of the magazine, when 2% tax is added to it.

(A) $1.53 (B) $1.65 (C) $1.72 (D) $0.03

12. The original price of a refrigerator is $15,000.00 and tax added to it is 6%. What is the sale price of the refrigerator ?

(A) $900.00 (B) $17,250.00 (C) $14,100.00 (D) $15,900.00

13. A box of cookies was on sale with 6% of tax added. Calculate the sale price of the box of cookies if the original price is \$6.25.

(A) \$7.50 (B) \$0.38 (C) \$6.63 (D) \$5.88

14. What is the selling price of a i-phone that is priced \$210.00 and with levied tax of 5% on the price ?

(A) \$199.50 (B) \$10.50 (C) \$220.50 (D) \$252.00

15. If the original price of yaztee game is \$39.99 and if 1% of tax is added to the price, calculate the selling price.

(A) \$0.40 (B) \$33.99 (C) \$40.39 (D) \$39.59

1. Find the percentage change from 13.6 to 18.

(A) 24.4% increase
(B) 4.4% decrease
(C) 132.4% increase
(D) 32.4% increase

2. Find the percentage change from 95 to 91.

(A) 14.4% increase
(B) 4.2% decrease
(C) 132.4% increase
(D) 3.4% increase

3. Find the percentage change from 33 to 60.

(A) 81.8% decrease
(B) 27% increase
(C) 81.8% increase
(D) 98.4% increase

4. Find the percentage change from 81.7 to 66.

(A) 23.8% increase
(B) 19.2% decrease
(C) 15.7% increase
(D) 80.8% decrease

5. Find the percentage change from 82 to 63.

(A) 19% decrease
(B) 30.2% increase
(C) 19% increase
(D) 23.2% decrease

6. Find the percentage change from 76 to 12.

(A) 15.8% decrease
(B) 533.3% increase
(C) 69.5% decrease
(D) 84.2% decrease

7. Find the percentage change from 50 to 99.

(A) 198% increase
(B) 49.5% decrease
(C) 98% increase
(D) 26.5% increase

8. Find the percentage change from 80.5 to 60.

(A) 121.3% increase
(B) 13% increase
(C) 25.47% decrease
(D) 117.6% increase

9. Find the percentage change from 85 to 43.

(A) 49.4% decrease

(B) 50.6% decrease

(C) 97.7% increase

(D) 97.7% decrease

10. Find the percentage change from 19 to 32.

(A) 68.4% increase

(B) 13% decrease

(C) 13% increase

(D) 168.4% increase

11. Find the percentage change from 24 to 37.

(A) 54.2% decrease

(B) 54.2% increase

(C) 7% decrease

(D) 35.1% decrease

12. Find the percentage change from 41 to 34.

(A) 7% decrease

(B) 17.1% increase

(C) 17.1% decrease

(D) 120.6% decrease

13. Find the percentage change from 31 to 5.3.

(A) 82.9% decrease
(B) 17.1% decrease
(C) 90.8% increase
(D) 25.7% increase

14. Find the percentage change from 47 to 8.

(A) 83% decrease
(B) 39% decrease
(C) 487.5% decrease
(D) 86.9% increase

15. Find the percentage change from 59 to 47.

(A) 12% increase
(B) 20.3% decrease
(C) 125.5% decrease
(D) 25.5% increase

Find the square root of the below.

1. $-\sqrt{\frac{36}{121}}$

(A) $-\frac{1}{6}$ (B) $-\frac{6}{11}$ (C) $-\frac{5}{7}$ (D) $-\frac{4}{7}$

2. $\sqrt{\frac{36}{169}}$

(A) $\frac{6}{13}$ (B) 3 (C) $\frac{13}{14}$ (D) $\frac{4}{7}$

3. $-\sqrt{\frac{144}{196}}$

(A) $-\frac{3}{5}$ (B) $-\frac{6}{7}$ (C) $-\frac{5}{6}$ (D) $-\frac{2}{7}$

4. $-\sqrt{\frac{1}{144}}$

(A) $-\frac{8}{11}$ (B) $-\frac{11}{13}$ (C) $-\frac{11}{14}$ (D) $-\frac{1}{12}$

Find the square root of the below.

5. $-\sqrt{\frac{4}{121}}$

(A) $-\frac{11}{13}$ (B) $-\frac{2}{11}$ (C) $-\frac{6}{7}$ (D) $-\frac{5}{6}$

6. $\sqrt{\frac{100}{121}}$

(A) $\frac{10}{11}$ (B) $\frac{1}{4}$ (C) $\frac{1}{8}$ (D) $\frac{3}{13}$

7. $-\sqrt{\frac{81}{100}}$

(A) $-\frac{6}{7}$ (B) $-\frac{9}{10}$ (C) -7 (D) $-\frac{6}{13}$

8. $-\sqrt{\frac{81}{4}}$

(A) $-\frac{7}{13}$ (B) $-\frac{5}{14}$ (C) $-4\frac{1}{2}$ (D) $-\frac{2}{3}$

Find the square root of the below.

9. $-\sqrt{\frac{1}{100}}$

(A) $-\frac{1}{6}$ (B) $-\frac{1}{10}$ (C) $-\frac{3}{4}$ (D) $-\frac{4}{11}$

10. $\sqrt{\frac{64}{81}}$

(A) $\frac{2}{3}$ (B) $\frac{1}{12}$ (C) $\frac{1}{2}$ (D) $\frac{8}{9}$

11. $-\sqrt{\frac{25}{64}}$

(A) $-\frac{2}{13}$ (B) $-\frac{1}{2}$ (C) $-\frac{5}{8}$ (D) $-\frac{2}{11}$

12. $\sqrt{\frac{169}{121}}$

(A) $\frac{1}{6}$ (B) $1\frac{2}{11}$ (C) $\frac{1}{2}$ (D) $\frac{3}{4}$

Find the square root of the below.

13. $\sqrt{\frac{64}{100}}$

(A) $\frac{1}{5}$ (B) $\frac{4}{5}$ (C) $\frac{2}{3}$ (D) $\frac{11}{14}$

14. $\sqrt{\frac{25}{121}}$

(A) $\frac{7}{13}$ (B) $\frac{2}{7}$ (C) $\frac{10}{11}$ (D) $\frac{5}{11}$

15. $\sqrt{\frac{1}{36}}$

(A) $1\frac{1}{6}$ (B) $\frac{1}{6}$ (C) $\frac{4}{11}$ (D) $\frac{7}{10}$

Simplify the below radicals.

1. $-2\sqrt{3} + 3\sqrt{3} + 2\sqrt{27}$

(A) $5\sqrt{3}$
(B) $7\sqrt{3}$
(C) $11\sqrt{3}$
(D) $14\sqrt{3}$

2. $2\sqrt{6} + 2\sqrt{5} + 3\sqrt{54}$

(A) $20\sqrt{6} + 6\sqrt{5}$
(B) $11\sqrt{6} + 4\sqrt{5}$
(C) $11\sqrt{6} + 6\sqrt{5}$
(D) $11\sqrt{6} + 2\sqrt{5}$

3. $-3\sqrt{5} + 3\sqrt{20} - \sqrt{6}$

(A) $-2\sqrt{6} + 6\sqrt{5}$
(B) $3\sqrt{5} - \sqrt{6}$
(C) $-\sqrt{6}$
(D) $-2\sqrt{6}$

4. $3\sqrt{5} - 2\sqrt{5} + 3\sqrt{6}$

(A) $\sqrt{5} + 3\sqrt{6}$
(B) $-\sqrt{5} + 3\sqrt{6}$
(C) $-3\sqrt{5} + 6\sqrt{6}$
(D) $-\sqrt{5} + 6\sqrt{6}$

Simplify the below radicals.

5. $-3\sqrt{5} + 2\sqrt{5} - \sqrt{5}$

(A) $-4\sqrt{5}$

(B) $-3\sqrt{3}$

(C) 0

(D) $-2\sqrt{5}$

6. $-2\sqrt{12} + 2\sqrt{3} - 2\sqrt{18}$

(A) $-2\sqrt{3} - 6\sqrt{2}$

(B) $-6\sqrt{2} + 4\sqrt{3}$

(C) $-6\sqrt{2} + 2\sqrt{3}$

(D) $-6\sqrt{2}$

7. $-\sqrt{24} + 3\sqrt{24} - 3\sqrt{2}$

(A) $-3\sqrt{2}$

(B) $2\sqrt{6} - 3\sqrt{2}$

(C) $4\sqrt{6} - 3\sqrt{2}$

(D) $-3\sqrt{2} - 2\sqrt{6}$

8. $-2\sqrt{27} - \sqrt{20} + 2\sqrt{27}$

(A) $-6\sqrt{5} + 6\sqrt{3}$

(B) $-4\sqrt{5} + 6\sqrt{3}$

(C) $-2\sqrt{5} + 6\sqrt{3}$

(D) $-2\sqrt{5}$

Simplify the below radicals.

9. $2\sqrt{2} - 3\sqrt{5} + 3\sqrt{18}$

(A) $11\sqrt{2} - 6\sqrt{5}$
(B) $11\sqrt{2} - 3\sqrt{5}$
(C) $20\sqrt{2} - 9\sqrt{5}$
(D) $11\sqrt{2} - 9\sqrt{5}$

10. $-3\sqrt{2} + 3\sqrt{2} + 3\sqrt{20}$

(A) $12\sqrt{5} - 3\sqrt{2}$
(B) $6\sqrt{5} + 3\sqrt{2}$
(C) $12\sqrt{5}$
(D) $6\sqrt{5}$

11. $3\sqrt{8} - 2\sqrt{2} + 3\sqrt{24}$

(A) $6\sqrt{6} - 2\sqrt{2}$
(B) $6\sqrt{6}$
(C) $4\sqrt{2} + 6\sqrt{6}$
(D) $2\sqrt{2} + 6\sqrt{6}$

12. $-3\sqrt{20} - 3\sqrt{5} - 2\sqrt{3}$

(A) $-4\sqrt{3}$
(B) $-4\sqrt{3} - 6\sqrt{5}$
(C) $-3\sqrt{5} - 2\sqrt{3}$
(D) $-9\sqrt{5} - 2\sqrt{3}$

Simplify the below radicals.

13. $3\sqrt{54} - \sqrt{8} + 3\sqrt{2}$

(A) $18\sqrt{6} + 7\sqrt{2}$

(B) $9\sqrt{6} + 4\sqrt{2}$

(C) $9\sqrt{6} + \sqrt{2}$

(D) $18\sqrt{6} + 4\sqrt{2}$

14. $-2\sqrt{18} + 2\sqrt{20} - 2\sqrt{8}$

(A) $-14\sqrt{2} + 4\sqrt{5}$

(B) $-10\sqrt{2} + 4\sqrt{5}$

(C) $-24\sqrt{2} + 4\sqrt{5}$

(D) $-20\sqrt{2} + 4\sqrt{5}$

15. $-3\sqrt{5} + 3\sqrt{6} - 2\sqrt{45}$

(A) $-9\sqrt{5} + 3\sqrt{6}$

(B) $-9\sqrt{5} + 6\sqrt{6}$

(C) $-15\sqrt{5} + 6\sqrt{6}$

(D) $-18\sqrt{5} + 6\sqrt{6}$

Simplify the below radicals.

1. $\sqrt{2}(\sqrt{5}+5\sqrt{3})$

(A) 35

(B) $5\sqrt{5}+5$

(C) $2\sqrt{5}+2$

(D) $\sqrt{10}+5\sqrt{6}$

2. $\sqrt{15}(\sqrt{5}+3)$

(A) $\sqrt{30}+2\sqrt{5}$

(B) $5\sqrt{3}+3\sqrt{15}$

(C) $5\sqrt{2}+2\sqrt{3}$

(D) $-20\sqrt{3}+\sqrt{30}$

3. $-\sqrt{5}(\sqrt{6}-\sqrt{10})$

(A) $2\sqrt{3}+5$

(B) $10\sqrt{2}+5\sqrt{3}$

(C) $\sqrt{30}+4$

(D) $-\sqrt{30}+5\sqrt{2}$

4. $\sqrt{10}(3\sqrt{10}+\sqrt{3})$

(A) $3\sqrt{3}+3$

(B) $-5\sqrt{30}+5\sqrt{2}$

(C) $30+\sqrt{30}$

(D) $3\sqrt{5}+2$

Simplify the below radicals.

5. $-2\sqrt{10}(4+4\sqrt{2})$

(A) $4\sqrt{3}+\sqrt{30}$

(B) $2\sqrt{5}+4$

(C) $2\sqrt{2}+3$

(D) $-8\sqrt{10}-16\sqrt{5}$

6. $\sqrt{6}(-5\sqrt{2}+3)$

(A) $-2\sqrt{30}+3$

(B) $-10\sqrt{3}+3\sqrt{6}$

(C) $-4\sqrt{5}+5\sqrt{2}$

(D) 12

7. $\sqrt{15}(\sqrt{3}+4)$

(A) $4\sqrt{3}+6\sqrt{5}$

(B) $3\sqrt{30}+5$

(C) -7

(D) $3\sqrt{5}+4\sqrt{15}$

8. $-3\sqrt{10}(4+2\sqrt{2})$

(A) $\sqrt{30}+3$

(B) $4+5\sqrt{2}$

(C) $-12\sqrt{10}-12\sqrt{5}$

(D) $4+12\sqrt{3}$

Simplify the below radicals.

9. $\sqrt{5}(-4\sqrt{10}+\sqrt{3})$

(A) $14\sqrt{3}$

(B) $15\sqrt{5}+5$

(C) $-2\sqrt{5}$

(D) $-20\sqrt{2}+\sqrt{15}$

10. $\sqrt{15}(3\sqrt{10}+\sqrt{3})$

(A) $-20\sqrt{3}-10\sqrt{2}$

(B) $2\sqrt{5}+2$

(C) $3\sqrt{3}+4\sqrt{2}$

(D) $15\sqrt{6}+3\sqrt{5}$

11. $\sqrt{6}(4\sqrt{3}+2)$

(A) 12

(B) $12\sqrt{2}+2\sqrt{6}$

(C) $3\sqrt{2}+15\sqrt{3}$

(D) $3\sqrt{2}+3$

12. $\sqrt{10}(-4\sqrt{2}+3)$

(A) $8\sqrt{3}+2\sqrt{5}$

(B) $-8\sqrt{5}+3\sqrt{10}$

(C) $3\sqrt{30}+3$

(D) $-6\sqrt{3}+3$

Simplify the below radicals.

13. $-\sqrt{5}(\sqrt{3}+3)$

(A) $-\sqrt{15}-3\sqrt{5}$

(B) $3\sqrt{5}+3$

(C) $5\sqrt{3}+5$

(D) $3\sqrt{3}+4$

14. $-3\sqrt{3}(2+5\sqrt{6})$

(A) $-6\sqrt{3}-45\sqrt{2}$

(B) 29

(C) $-6\sqrt{2}+2$

(D) $12\sqrt{5}+4$

15. $-2\sqrt{15}(3-4\sqrt{5})$

(A) $-6\sqrt{15}+40\sqrt{3}$

(B) $6\sqrt{5}+3$

(C) 10

(D) $-25\sqrt{3}-6\sqrt{5}$

Simplify the below radicals.

1. $$\frac{3}{2+3\sqrt{5}}$$

(A) $\frac{2+3\sqrt{5}}{3}$

(B) $\frac{\sqrt{5}+1}{3}$

(C) $\frac{-15+5\sqrt{5}}{4}$

(D) $\frac{-6+9\sqrt{5}}{41}$

2. $$\frac{3}{4\sqrt{5}+\sqrt{2}}$$

(A) $\frac{6-2\sqrt{2}}{7}$

(B) $\frac{-5+\sqrt{5}}{4}$

(C) $\frac{4\sqrt{5}-\sqrt{2}}{26}$

(D) $\frac{-9+15\sqrt{2}}{41}$

3. $$\frac{5}{2(\sqrt{7}-\sqrt{2})}$$

(A) $\frac{\sqrt{7}+\sqrt{2}}{2}$

(B) $\frac{20\sqrt{7}-15\sqrt{2}}{78}$

(C) $\frac{15+3\sqrt{7}}{23}$

(D) $4-2\sqrt{2}$

Simplify the below radicals.

4. $$\frac{4}{\sqrt{3}+4\sqrt{5}}$$

(A) $\frac{-4\sqrt{3} - 16}{13}$

(B) $\frac{3+3\sqrt{5}}{4}$

(C) $-8\sqrt{5} - 4\sqrt{15}$

(D) $\frac{-4\sqrt{3} + 16\sqrt{5}}{77}$

5. $$\frac{2}{\sqrt{2}+\sqrt{5}}$$

(A) $\frac{15 - 3\sqrt{3}}{22}$

(B) $\frac{-\sqrt{10} - 5\sqrt{5}}{23}$

(C) $\frac{3 + 3\sqrt{3}}{4}$

(D) $\frac{-2\sqrt{2} + 2\sqrt{5}}{3}$

6. $$\frac{3}{2+2\sqrt{3}}$$

(A) $3 + \sqrt{5}$

(B) $\frac{2 + 2\sqrt{3}}{3}$

(C) $\frac{-3 + 3\sqrt{3}}{4}$

(D) $\frac{9 - 3\sqrt{5}}{4}$

Simplify the below radicals.

7. $\dfrac{\sqrt{5}}{3\sqrt{5}-4\sqrt{2}}$

(A) $-\sqrt{6}+3$

(B) $\dfrac{15+4\sqrt{10}}{13}$

(C) $\dfrac{1-2\sqrt{3}}{11}$

(D) $\dfrac{-4+4\sqrt{2}}{5}$

8. $\dfrac{2}{2+\sqrt{2}}$

(A) $\dfrac{2+\sqrt{2}}{2}$

(B) $\dfrac{8-6\sqrt{5}}{29}$

(C) $\dfrac{-20+5\sqrt{5}}{11}$

(D) $2-\sqrt{2}$

9. $\dfrac{2}{-4-3\sqrt{3}}$

(A) $\dfrac{8-6\sqrt{3}}{11}$

(B) $\dfrac{-2\sqrt{3}-8}{13}$

(C) $\dfrac{3+5\sqrt{5}}{58}$

(D) $\dfrac{-1-\sqrt{2}}{2}$

Simplify the below radicals.

10. $$\frac{2}{-3+4\sqrt{3}}$$

(A) $\frac{-1-\sqrt{3}}{2}$

(B) $\frac{6+8\sqrt{3}}{39}$

(C) $\frac{10-15\sqrt{5}}{41}$

(D) $\frac{-2+\sqrt{3}}{2}$

11. $$\frac{3}{-4-\sqrt{5}}$$

(A) $\frac{-5-10\sqrt{2}}{14}$

(B) $\frac{-4-\sqrt{5}}{5}$

(C) $\frac{-12+3\sqrt{5}}{11}$

(D) $\frac{10+2\sqrt{3}}{11}$

12. $$\frac{4\sqrt{3}}{\sqrt{5}+4}$$

(A) $\frac{-4\sqrt{15}+16\sqrt{3}}{11}$

(B) $\frac{8-10\sqrt{2}}{17}$

(C) $\frac{1-\sqrt{5}}{6}$

(D) $\frac{4-3\sqrt{3}}{5}$

Simplify the below radicals.

13. $\dfrac{2}{\sqrt{5}+4\sqrt{3}}$

(A) $\dfrac{3\sqrt{2}+\sqrt{5}}{3}$

(B) $\dfrac{3-\sqrt{2}}{4}$

(C) $\dfrac{-2\sqrt{5}+8\sqrt{3}}{43}$

(D) $-1-\sqrt{3}$

14. $\dfrac{3}{2+5\sqrt{5}}$

(A) $\dfrac{-20+16\sqrt{5}}{55}$

(B) $\dfrac{2+5\sqrt{5}}{3}$

(C) $\dfrac{6-3\sqrt{2}}{2}$

(D) $\dfrac{-6+15\sqrt{5}}{121}$

15. $\dfrac{4}{-4-\sqrt{3}}$

(A) $\dfrac{-16+4\sqrt{3}}{13}$

(B) $\dfrac{5-4\sqrt{5}}{2}$

(C) $\dfrac{-4-\sqrt{3}}{4}$

(D) $\dfrac{25-10\sqrt{3}}{13}$

Evaluate the below and find the value of the variable.

1. $8 = \sqrt{2b + 60}$

(A) 2 (B) −4 (C) −9 (D) 3

2. $-9 = -9\sqrt{\frac{k}{5}}$

(A) 5 (B) 6, 2 (C) 6 (D) 2

3. $8 = 8\sqrt{-6 - x}$

(A) −9, 7 (B) 7 (C) −4 (D) −7

4. $-2 = -4 + \sqrt{2x}$

(A) −1 (B) 2 (C) −10, 2 (D) -2,2

Evaluate the below and find the value of the variable.

5. $\sqrt{n+5} = 8$

(A) 59 (B) 9 (C) 59, 4 (D) −3, 4

6. $14 = \sqrt{5m} + 9$

(A) 2 (B) −5, 8 (C) 5 (D) -5

7. $9 = 9\sqrt{m+2}$

(A) 1 (B) −1, 9 (C) 1, 9 (D) −1

8. $\sqrt{p+8} + 6 = 15$

(A) 73, 10 (B) 73 (C) 6 (D) −73

Evaluate the below and find the value of the variable.

9. $\sqrt{16a} + 5 = 9$

(A) –1 (B) –3 (C) 1 (D) –1, –2

10. $11 = \sqrt{b + 10} + 8$

(A) –1 (B) 6 (C) 7 (D) 1, –7

11. $\sqrt{4 - 10x} - 2 = 0$

(A) –2, 2 (B) 0 (C) 2 (D) –2

12. $5 + \sqrt{\frac{r}{5}} = 5$

(A) 2 (B) 2,0 (C) –8, 0 (D) 0

Evaluate the below and find the value of the variable.

13.

$$\sqrt{9x} - 8 = -5$$

(A) 1, -1 (B) -2, -1 (C) 1 (D) 9

14.

$$8 + \sqrt{p + 5} = 15$$

(A) -44 (B) 44, -44 (C) 44 (D) -8

15.

$$\sqrt{\frac{m}{9}} + 6 = 7$$

(A) 9 (B) -9 (C) 5, -9 (D) -7

Simplify the below radical to the lowest terms

1. $\sqrt{245n}$

(A) $6n^2\sqrt{7}$ (B) $7\sqrt{5n}$ (C) $8\sqrt{6n}$ (D) $10n$

2. $\sqrt{12r^3}$

(A) $4\sqrt{5r}$ (B) $10\sqrt{r}$ (C) $2r\sqrt{3r}$ (D) $4r$

3. $\sqrt{72a}$

(A) $7a\sqrt{7a}$ (B) $2a\sqrt{3}$ (C) $3a\sqrt{7a}$ (D) $6\sqrt{2a}$

4. $\sqrt{18x}$

(A) $3\sqrt{2x}$ (B) $7x\sqrt{7x}$ (C) $12\sqrt{x}$ (D) $10x\sqrt{2x}$

Simplify the below radical to the lowest terms

5. $\sqrt{512n^3}$

(A) $16n\sqrt{2n}$ (B) $5n^2\sqrt{3}$ (C) $8n$ (D) $18n\sqrt{2n}$

6. $\sqrt{245x^2}$

(A) $7x\sqrt{5}$ (B) $8x\sqrt{3}$ (C) $3x\sqrt{6x}$ (D) $10x\sqrt{x}$

7. $\sqrt{24p^3}$

(A) $5p\sqrt{3p}$ (B) $7\sqrt{5p}$ (C) $2p\sqrt{6p}$ (D) $8p^2\sqrt{5}$

8. $\sqrt{108v^2}$

(A) $16v^2$ (B) $6v\sqrt{3}$ (C) $7\sqrt{2v}$ (D) $3v\sqrt{3}$

Simplify the below radical to the lowest terms

9. $\sqrt{448n^4}$

(A) $5n^2\sqrt{7}$ (B) $16n^2$ (C) $4n^2\sqrt{5}$ (D) $8n^2\sqrt{7}$

10. $\sqrt{64x^2}$

(A) $4x^2\sqrt{2}$ (B) $8x$ (C) $4x\sqrt{2x}$ (D) $8\sqrt{x}$

11. $\sqrt{36x}$

(A) $5x\sqrt{6}$ (B) $8x\sqrt{2x}$ (C) $6\sqrt{x}$ (D) $6x\sqrt{2}$

12. $\sqrt{24m^4}$

(A) $4m$ (B) $2m\sqrt{6m}$ (C) $6m\sqrt{7m}$ (D) $2m^2\sqrt{6}$

Simplify the below radical to the lowest terms

13. $\sqrt{80x}$

(A) $14x$ (B) $6x\sqrt{7x}$ (C) $7\sqrt{5x}$ (D) $4\sqrt{5x}$

14. $\sqrt{80v^4}$

(A) $2\sqrt{2v}$ (B) $6\sqrt{7v}$ (C) $4v^2\sqrt{5}$ (D) $7v^2\sqrt{3}$

15. $\sqrt{63x^2}$

(A) $3x\sqrt{7}$ (B) $4\sqrt{3x}$ (C) $16\sqrt{2x}$ (D) $5x^2\sqrt{6}$

Find the area of the below triangles.

1)

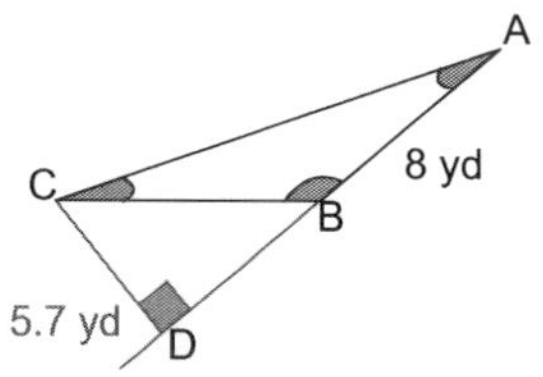

A) 18.3 yd² B) 11.4 yd²
C) 22.8 yd² D) 45.6 yd²

2)

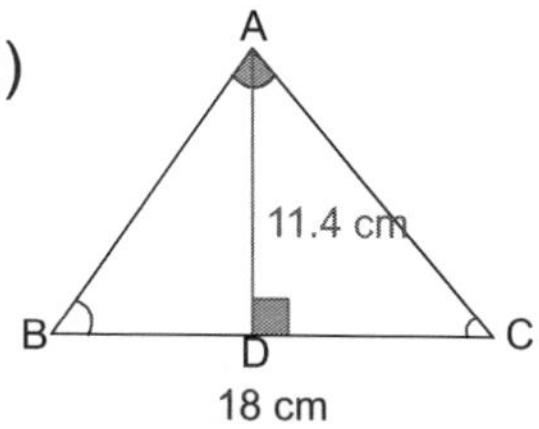

A) 205.2 cm² B) 51.3 cm²
C) 102.6 cm² D) 104.8 cm²

3)

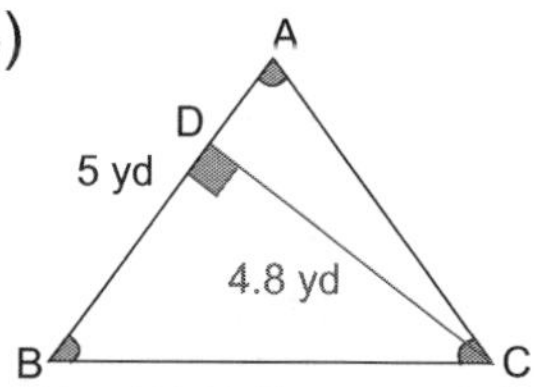

A) 12 yd² B) 16.3 yd²
C) 6 yd² D) 24 yd²

4)

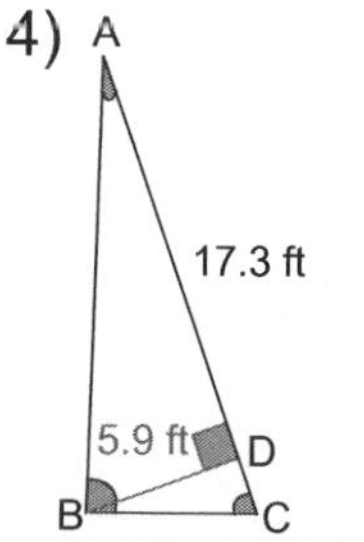

A) 25.5 ft² B) 56.635 ft²
C) 102.07 ft² D) 51.035 ft²

5)

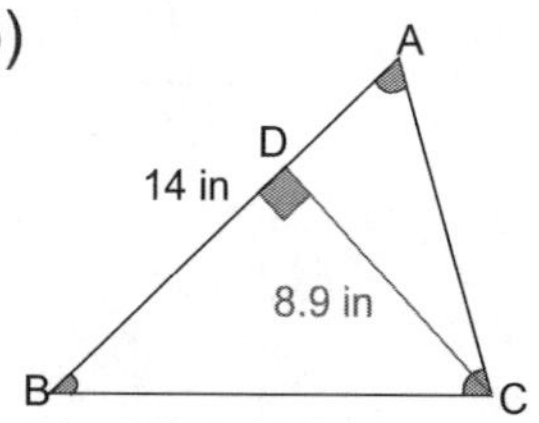

A) 62.3 in² B) 124.6 in²
C) 31.2 in² D) 67.1 in²

6)

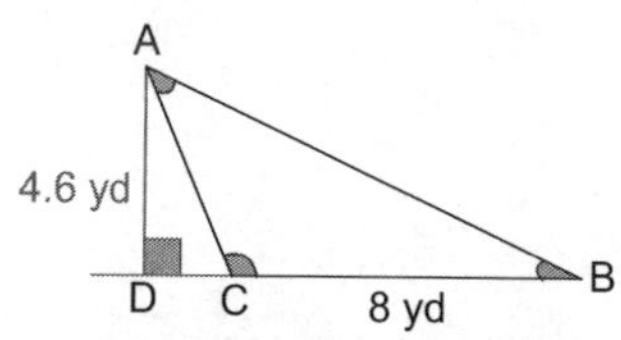

A) 36.8 yd² B) 14.1 yd²
C) 18.4 yd² D) 9.2 yd²

7)

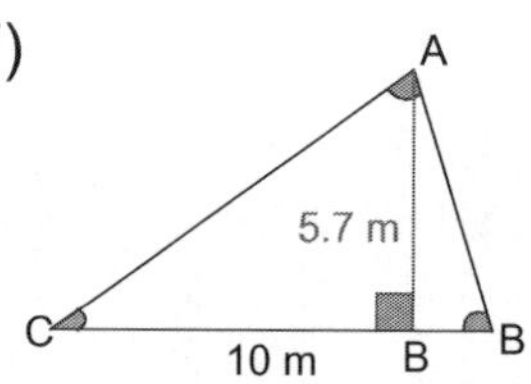

A) 14.3 m² B) 28.5 m²
C) 32.2 m² D) 57 m²

8)

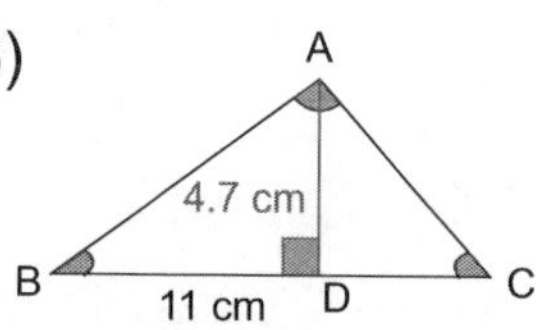

A) 12.9 cm² B) 51.7 cm²
C) 25.85 cm² D) 15.95 cm²

9)

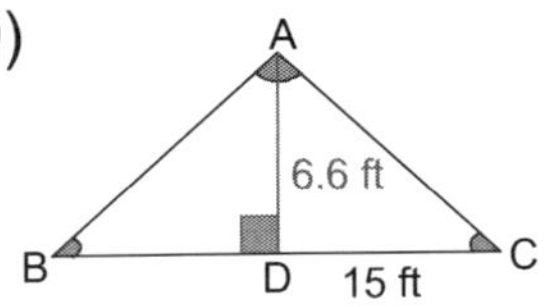

A) 43.6 ft²
B) 49.5 ft²
C) 99 ft²
D) 24.8 ft²

10)

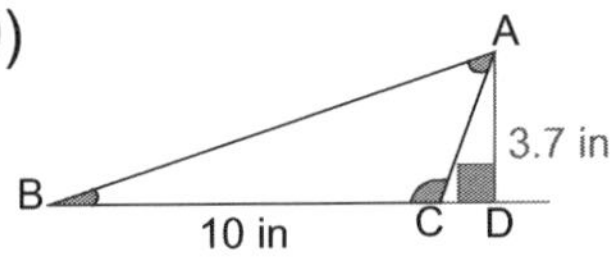

A) 37 in²
B) 18.5 in²
C) 30.4 in²
D) 15.2 in²

11)

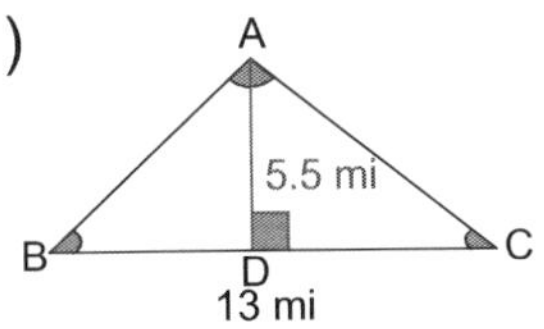

A) 38.45 mi²
B) 71.5 mi²
C) 35.75 mi²
D) 19.2 mi²

12)

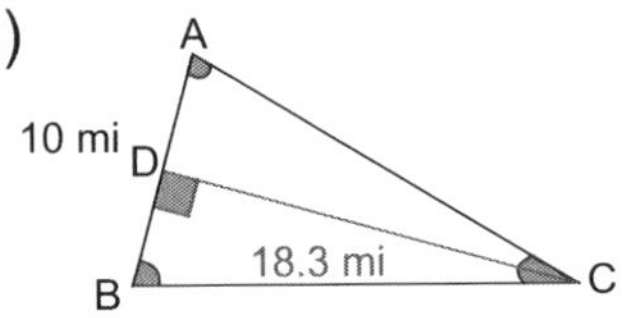

A) 45.8 mi²
B) 183 mi²
C) 100.1 mi²
D) 91.5 mi²

13)

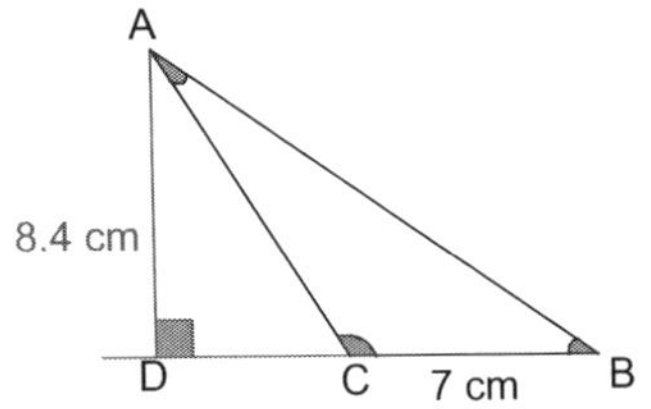

A) 14.7 cm²
B) 58.8 cm²
C) 36.5 cm²
D) 29.4 cm²

14)

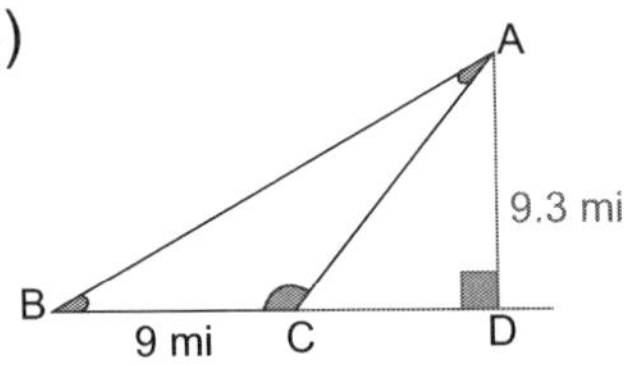

A) 20.9 mi²
B) 41.85 mi²
C) 83.7 mi²
D) 45.85 mi²

15)

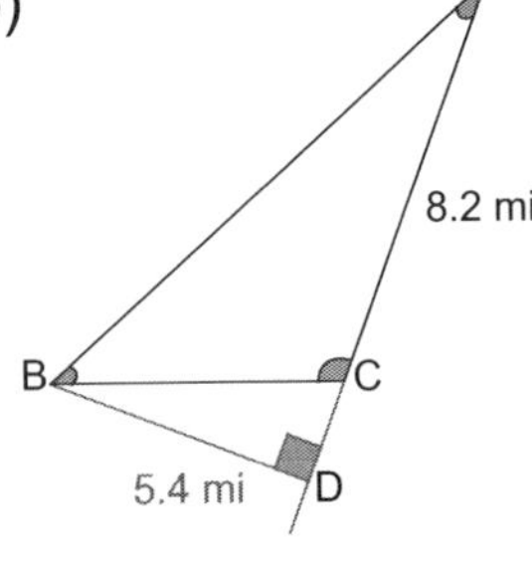

A) 11.1 mi²
B) 44.28 mi²
C) 28.44 mi²
D) 22.14 mi²

Identify the below quadrilateral and find its area.

1) 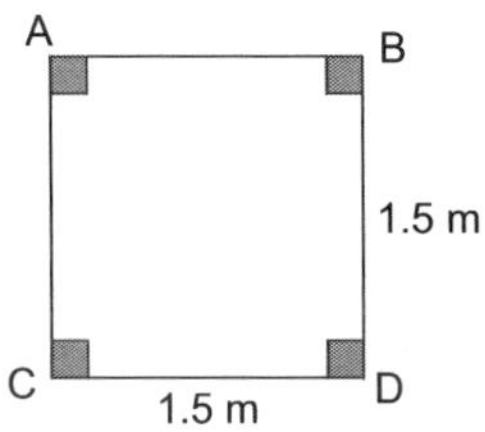

A) 1.1 m²
B) 8.65 m²
C) 4.5 m²
D) 2.25 m²

2) 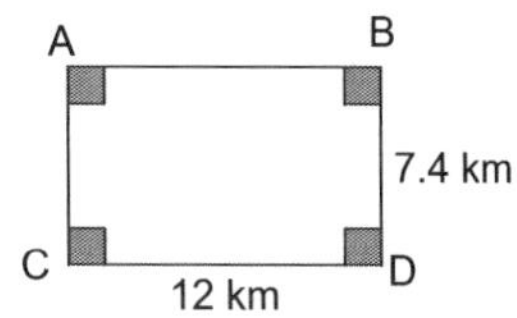

A) 177.6 km²
B) 44.4 km²
C) 88.8 km²
D) 82.1 km²

3) 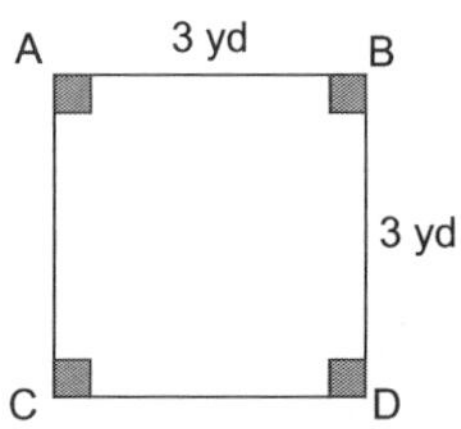

A) 18.3 yd²
B) 9 yd²
C) 4.5 yd²
D) 18 yd²

4) 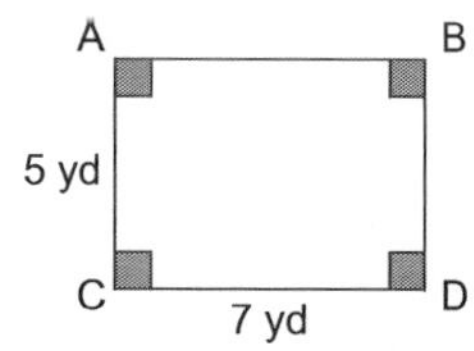

A) 70 yd²
B) 32 yd²
C) 17.5 yd²
D) 35 yd²

5) 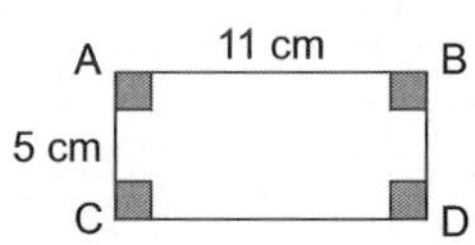

A) 110 cm²
B) 55 cm²
C) 60.7 cm²
D) 27.5 cm²

6) 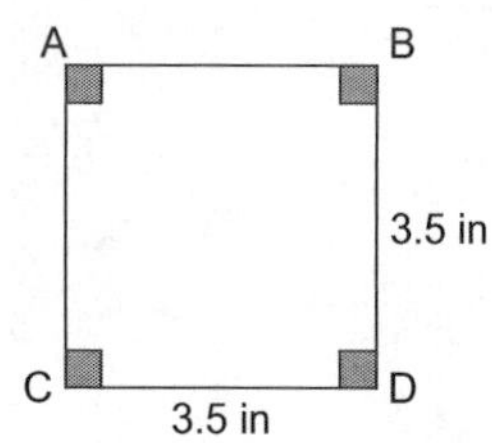

A) 13.65 in²
B) 24.5 in²
C) 12.25 in²
D) 6.1 in²

7) 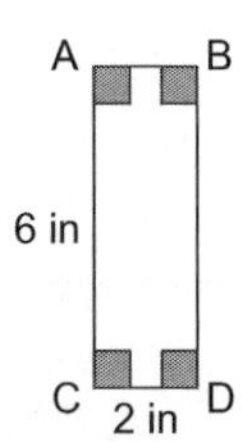

A) 6 in²
B) 5.5 in²
C) 24 in²
D) 12 in²

8) 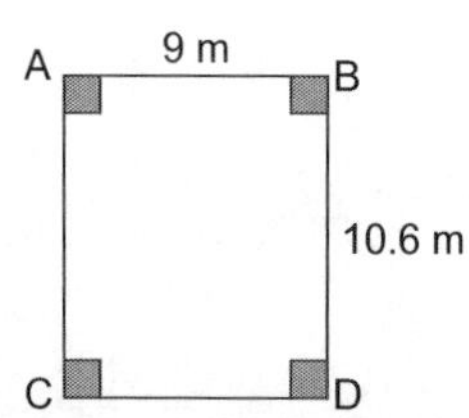

A) 107.1 m²
B) 95.4 m²
C) 47.7 m²
D) 98.9 m²

9)

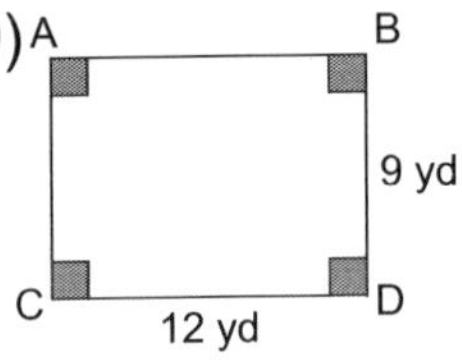

A) 108 yd² B) 54 yd²
C) 117.8 yd² D) 216 yd²

10)

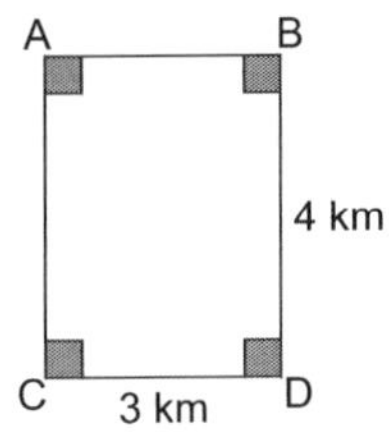

A) 24 km² B) 6 km²
C) 19.4 km² D) 12 km²

11)

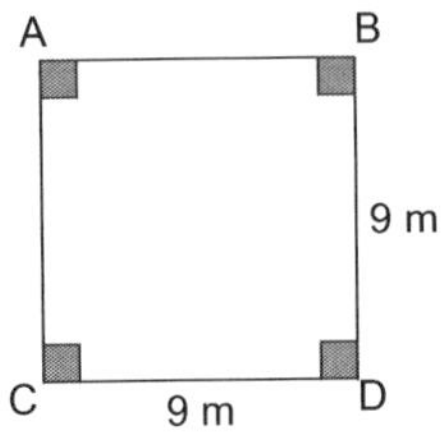

A) 40.5 m² B) 162 m²
C) 80.2 m² D) 81 m²

12)

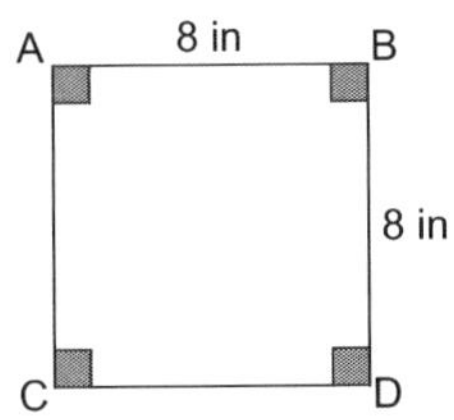

A) 128 in² B) 58.8 in²
C) 64 in² D) 32 in²

13)

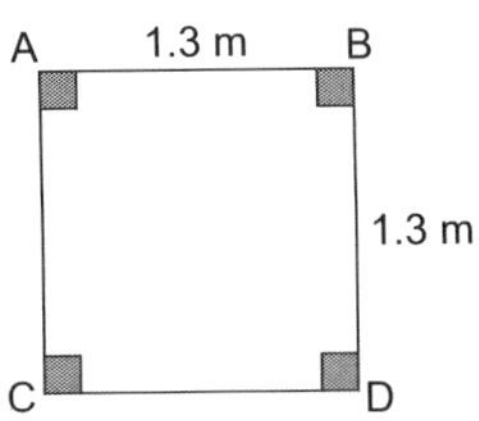

A) 3.38 m² B) 1.69 m²
C) 7.29 m² D) 0.8 m²

14)

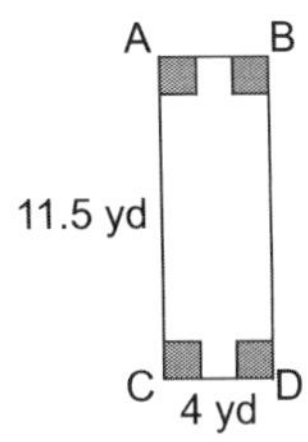

A) 48.2 yd² B) 92 yd²
C) 23 yd² D) 46 yd²

15)

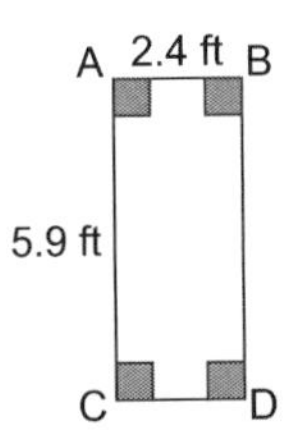

A) 14.16 ft² B) 7.1 ft²
C) 9.66 ft² D) 28.32 ft²

Identify the below quadrilateral and find its area.

1)

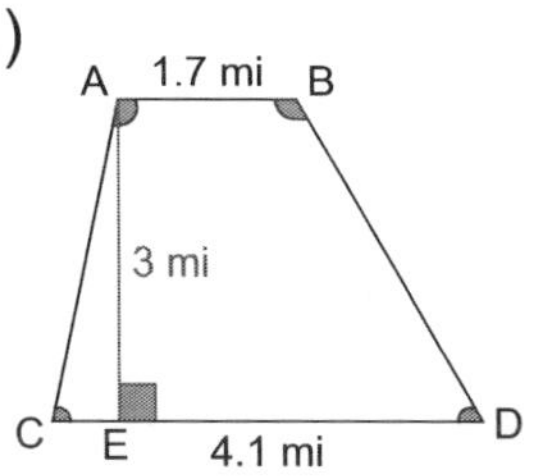

A) 16 mi²　B) 6.8 mi²
C) 8.7 mi²　D) 17.4 mi²

2)

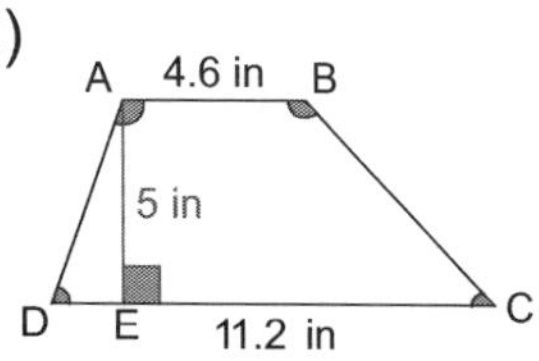

A) 36.4 in²　B) 79 in²
C) 39.5 in²　D) 19.8 in²

3)

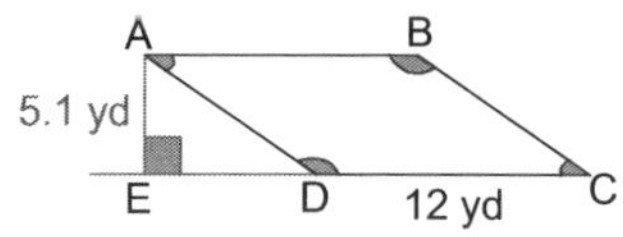

A) 122.4 yd²　B) 30.6 yd²
C) 70.1 yd²　D) 61.2 yd²

4)

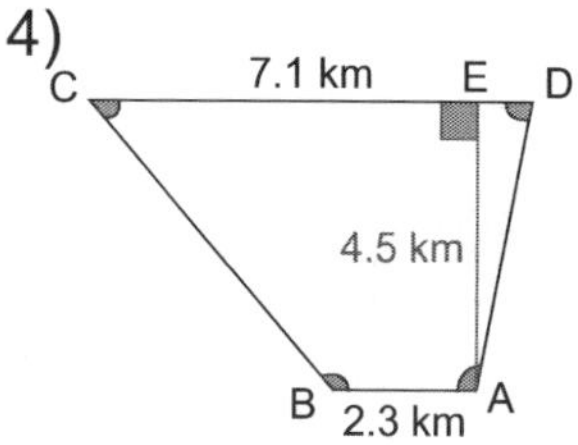

A) 21.15 km²　B) 42.3 km²
C) 10.6 km²　D) 30.95 km²

5)

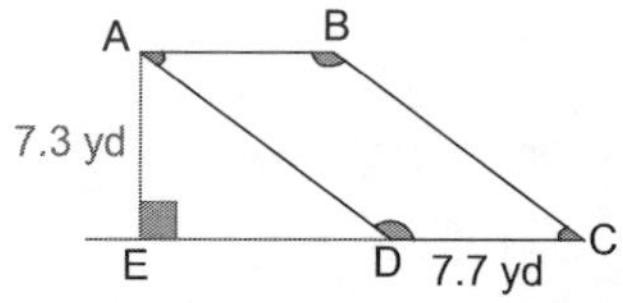

A) 28.1 yd²　B) 64.91 yd²
C) 56.21 yd²　D) 112.42 yd²

6)

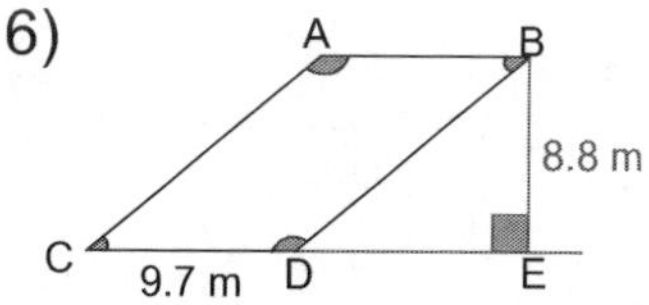

A) 85.36 m²　B) 170.72 m²
C) 91.26 m²　D) 42.7 m²

7)

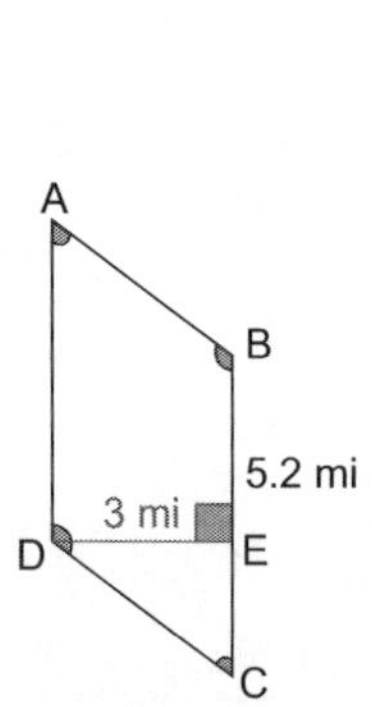

A) 15.6 mi²　B) 7.8 mi²
C) 31.2 mi²　D) 16.6 mi²

8)

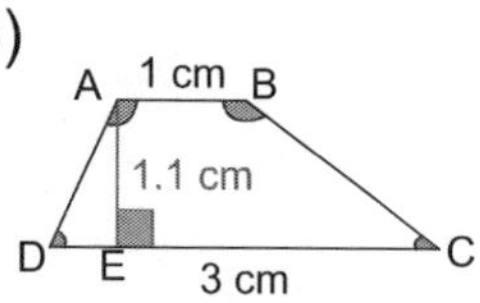

A) 4.4 cm²　B) 10.4 cm²
C) 1.1 cm²　D) 2.2 cm²

9) 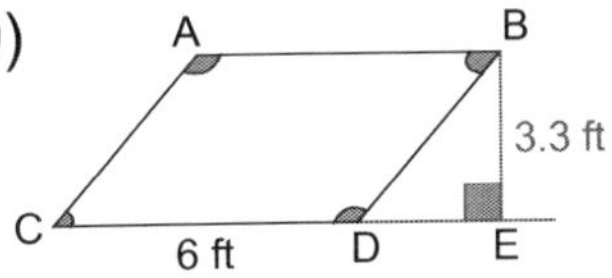

A) 19.8 ft² B) 14.7 ft²
C) 39.6 ft² D) 9.9 ft²

10) 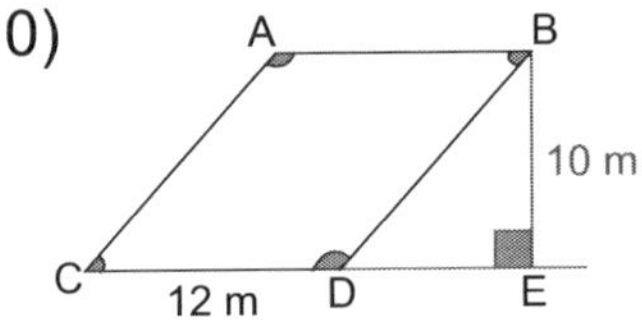

A) 120 m² B) 112.7 m²
C) 60 m² D) 240 m²

11) 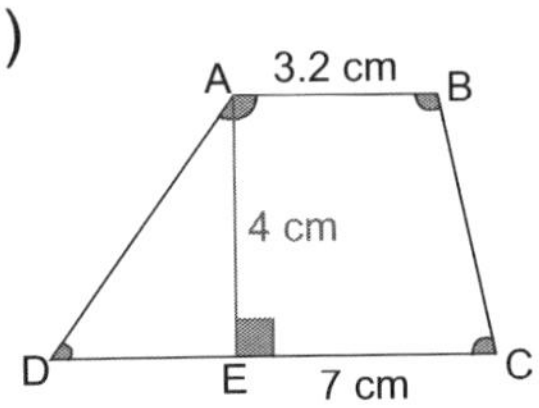

A) 40.8 cm² B) 10.2 cm²
C) 20.4 cm² D) 18.2 cm²

12) 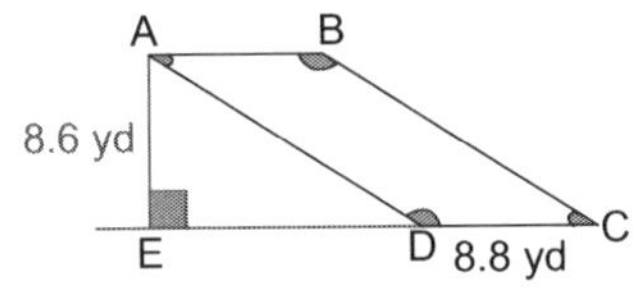

A) 36.2 yd² B) 144.96 yd²
C) 75.68 yd² D) 72.48 yd²

13) 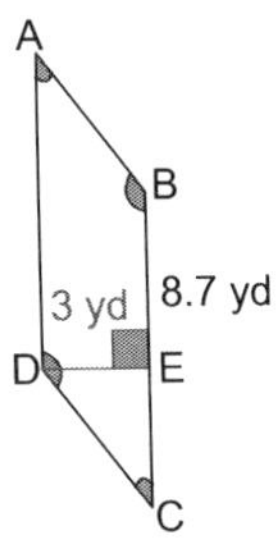

A) 26.1 yd² B) 52.2 yd²
C) 25.3 yd² D) 13.1 yd²

14) 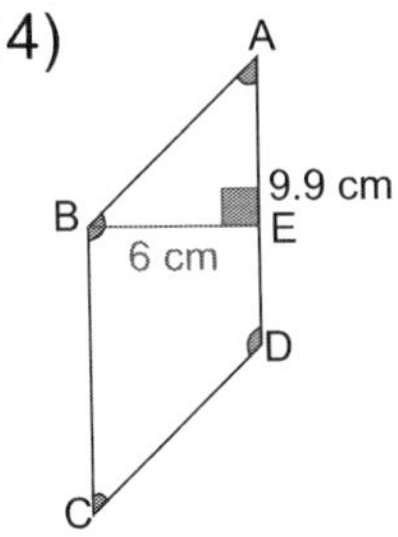

A) 50.7 cm² B) 118.8 cm²
C) 29.7 cm² D) 59.4 cm²

15) 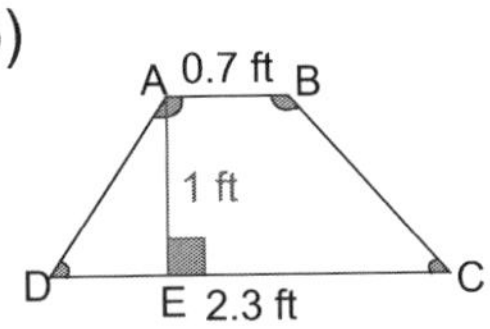

A) 1.5 ft² B) 5.2 ft²
C) 0.8 ft² D) 3 ft²

1. Find the area of the circle given diameter = 4 m.

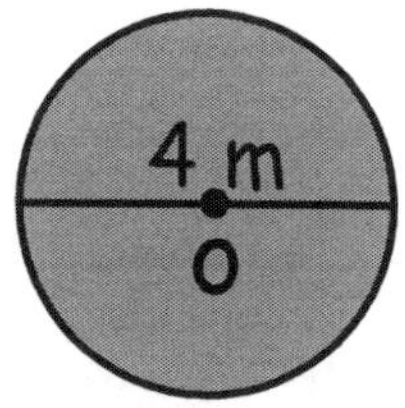

(A) $12\pi\ m^2$ (B) $9\pi\ m^2$

(C) $7\pi\ m^2$ (D) $4\pi\ m^2$

2. Find the area of the circle given diameter = 10 m.

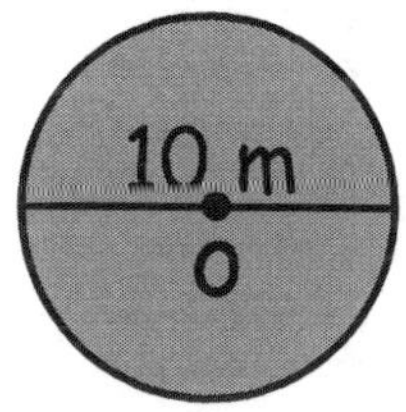

(A) $25\pi\ m^2$ (B) $50\pi\ m^2$

(C) $42\pi\ m^2$ (D) $34\pi\ m^2$

3. Find the area of the circle given diameter = 6 in.

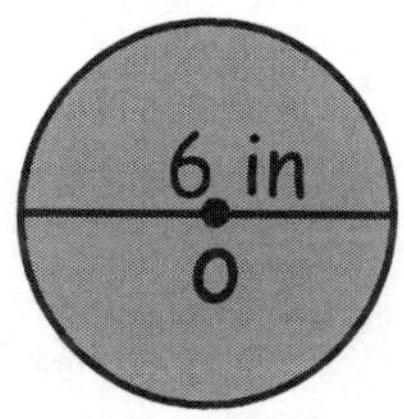

(A) $4\pi\ in^2$ (B) $12\pi\ in^2$

(C) $16\pi\ in^2$ (D) $9\pi\ in^2$

4. Find the area of the circle given diameter = 24 ft.

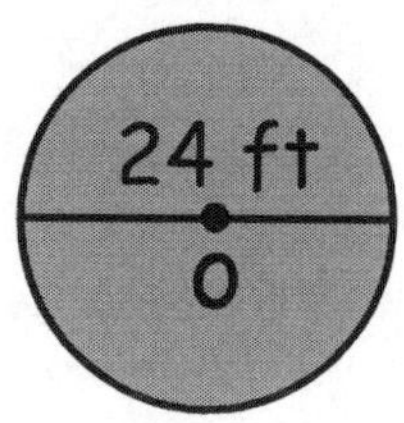

(A) $1296\pi\ ft^2$ (B) $144\pi\ ft^2$

(C) $36\pi\ ft^2$ (D) $9\pi\ ft^2$

5. Find the area of the circle given diameter = 12 in

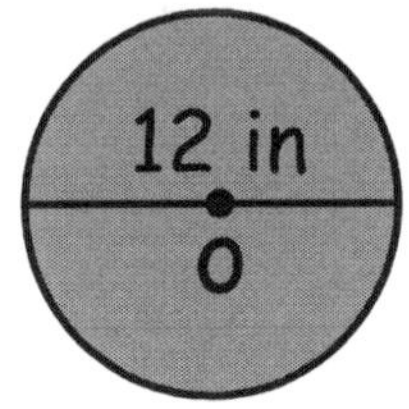

(A) $36\pi\ in^2$ (B) $62\pi\ in^2$

(C) $44\pi\ in^2$ (D) $52\pi\ in^2$

6. Find the area of the circle given diameter = 22 yd.

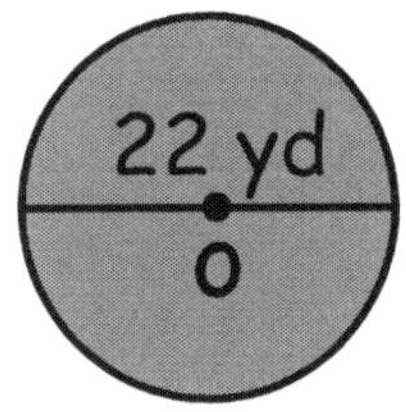

(A) $64\pi\ yd^2$ (B) $16\pi\ yd^2$

(C) $121\pi\ yd^2$ (D) $22\pi\ yd^2$

7. Find the area of the circle given diameter = 14 km.

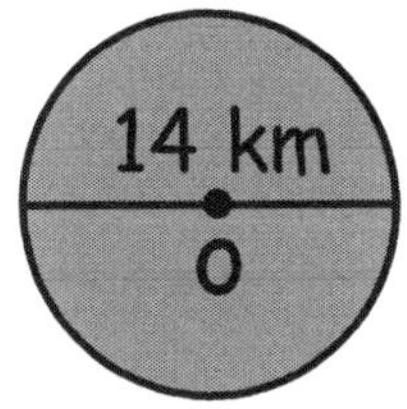

(A) $64\pi\ km^2$ (B) $8\pi\ km^2$

(C) $49\pi\ km^2$ (D) $59\pi\ km^2$

8. Find the area of the circle given diameter = 8 ft.

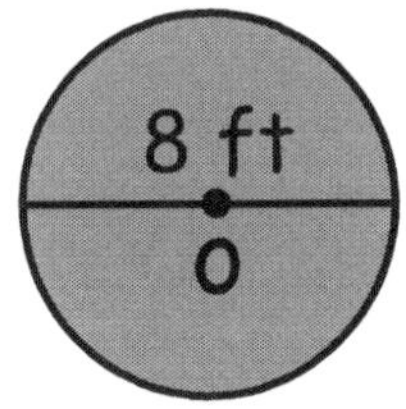

(A) $22\pi\ ft^2$ (B) $9\pi\ ft^2$

(C) $16\pi\ ft^2$ (D) $6\pi\ ft^2$

9. Find the area of the circle given diameter = 20 yd.

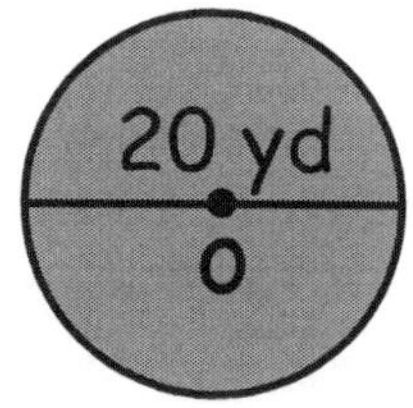

(A) 108π yd^2 (B) 100π yd^2

(C) 121π yd^2 (D) 102π yd^2

10. Find the area of the circle given diameter = 18 cm.

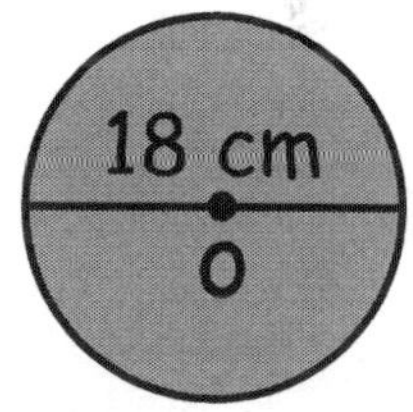

(A) 81π cm^2 (B) 372π cm^2

(C) 324π cm^2 (D) 339π cm^2

11. Find the area of the circle given diameter = 16 mi.

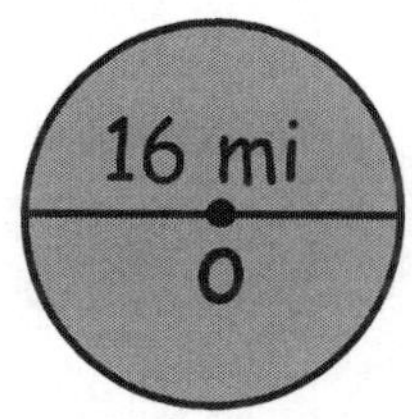

(A) 64π mi^2 (B) 8π mi^2

(C) 12π mi^2 (D) 13π mi^2

12. Find the circumference of the circle given radius = 7.6 in and round the answer to the nearest tenth.

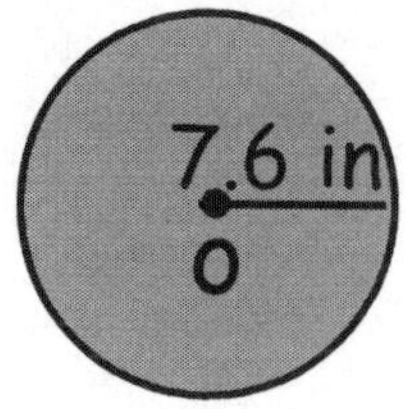

(A) 28.7 in (B) 17.3 in

(C) 23 in (D) 47.7 in

13. Find the circumference of the circle given radius = 3.1 ft and round the answer to the nearest tenth.

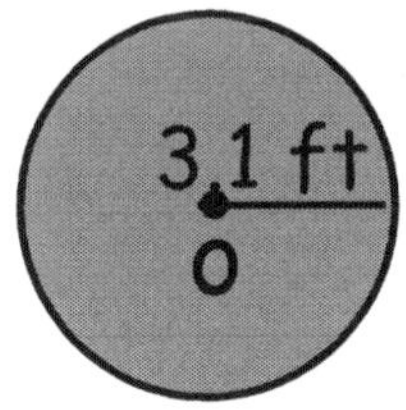

(A) 19.5 ft (B) 18.8 ft

(C) 21.4 ft (D) 19.1 ft

14. Find the circumference of the circle given radius = 10.1 yd and round the answer to the nearest tenth.

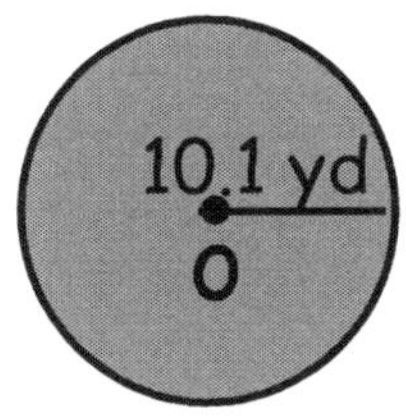

(A) 64.8 yd (B) 72.3 yd

(C) 67.3 yd (D) 63.4 yd

15. Find the circumference of the circle given radius = 8.8 mi and round the answer to the nearest tenth.

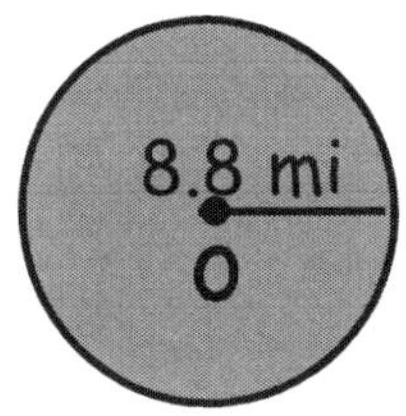

(A) 34 mi (B) 184 mi

(C) 55.3 mi (D) 27.7 mi

1. Find the volume of the sphere with a radius of 13.2 mi and round the answer to the nearest tenth.

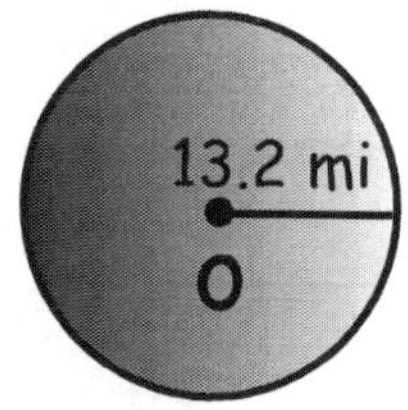

(A) 9634.1 mi^3 (B) 5587.7 mi^3

(C) 6705.2 mi^3 (D) 4817 mi^3

2. Find the volume of the sphere with a diameter of 13.6 mi and round the answer to the nearest tenth.

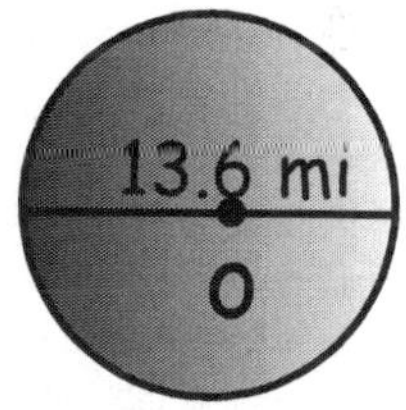

(A) 2370.8 mi^3 (B) 1185.4 mi^3

(C) 2679 mi^3 (D) 1317.1 mi^3

3. Find the volume of the sphere with a diameter of 12.6 m and round the answer to the nearest tenth.

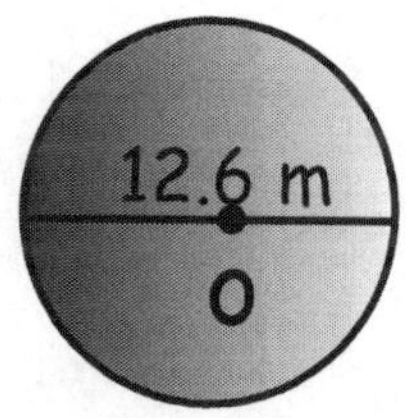

(A) 1149.9 m^3 (B) 1047.4 m^3

(C) 999.9 m^3 (D) 900.8 mi^3

4. Find the volume of the sphere with a radius of 12 ft and round the answer to the nearest tenth.

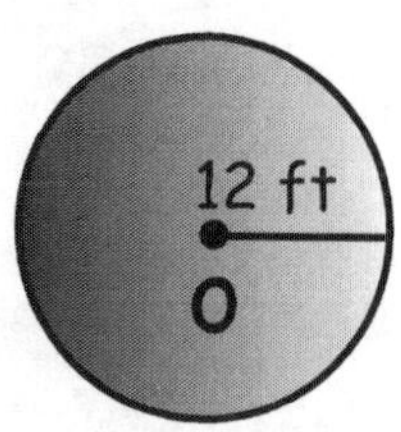

(A) 7238.2 ft^3 (B) 4620.9 ft^3

(C) 3619.1 ft^3 (D) 4125.8 ft^3

5. Find the volume of the sphere with a diameter of 24.6 mi and round the answer to the nearest tenth.

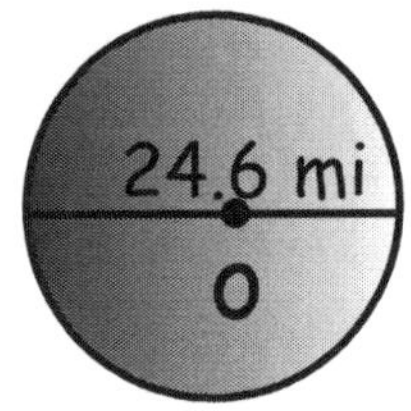

(A) 5827.4 mi^3 (B) 5244.7 mi^3

(C) 7794.8 mi^3 (D) 6547.6 mi^3

6. Find the volume of the sphere with a diameter of 2.2 ft and round the answer to the nearest tenth.

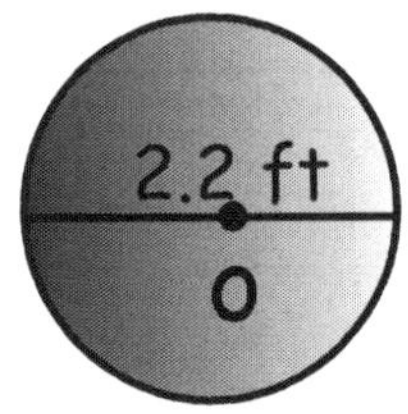

(A) 6.5 ft^3 (B) 13 ft^3

(C) 11.4 ft^3 (D) 5.6 ft^3

7. Find the volume of the sphere with a radius of 2.1 cm and round the answer to the nearest tenth.

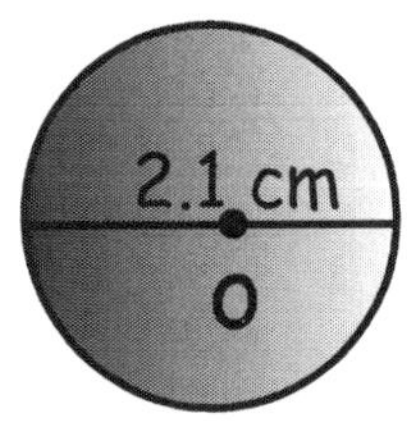

(A) 30.5 cm^3 (B) 38.8 cm^3

(C) 35 cm^3 (D) 43.8 cm^3

8. Find the volume of the sphere with a radius of 17.2 cm and round the answer to the nearest tenth.

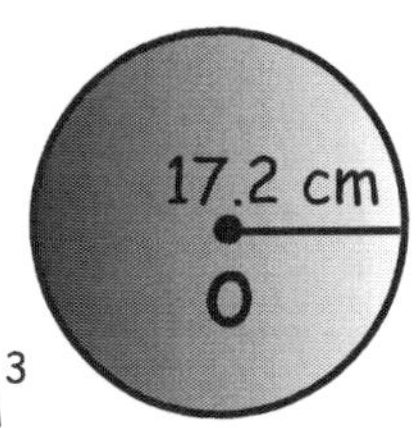

(A) 21314.1 cm^3 (B) 23659.1 cm^3

(C) 42628.9 cm^3 (D) 21314.4 cm^3

9. Find the volume of the sphere with a diameter of 28.6 km and round the answer to the nearest tenth.

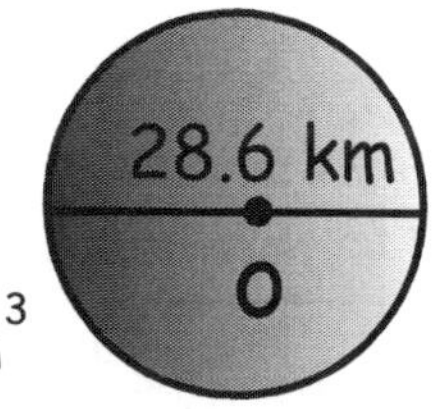

(A) 12248.9 km^3 (B) 10901.5 km^3

(C) 21803 km^3 (D) 23983.3 km^3

10. Find the volume of the sphere with a diameter of 4.8 cm and round the answer to the nearest tenth.

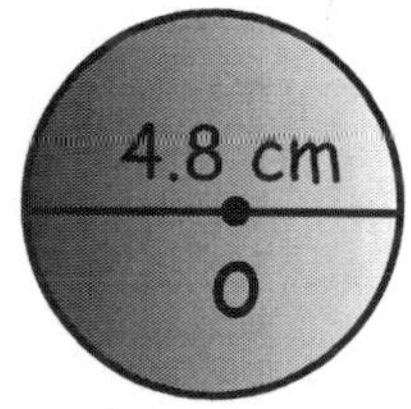

(A) 77 cm^3 (B) 48.1 cm^3

(C) 57.9 cm^3 (D) 96.2 cm^3

11. Find the volume of the sphere with a radius of 14.7 cm and round the answer to the nearest tenth.

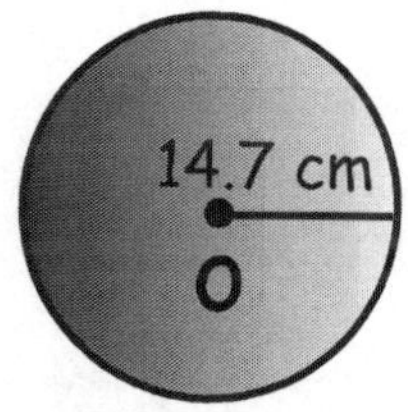

(A) 13305.8 cm^3 (B) 3729.5 cm^3

(C) 7917 cm^3 (D) 6652.9 cm^3

12. Find the volume of the sphere with a radius of 9.3 km and round the answer to the nearest tenth.

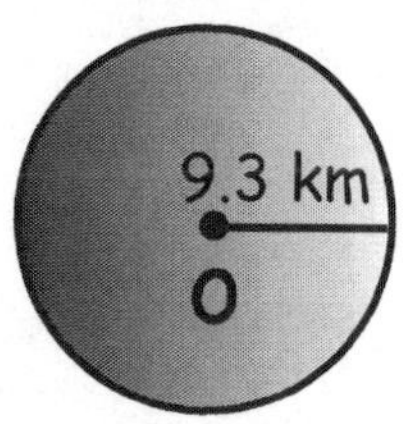

(A) 7816.8 km^3 (B) 3369.3 km^3

(C) 9380.2 km^3 (D) 3908.4 km^3

13. Find the volume of the sphere with a radius of 9 ft and round the answer to the nearest tenth.

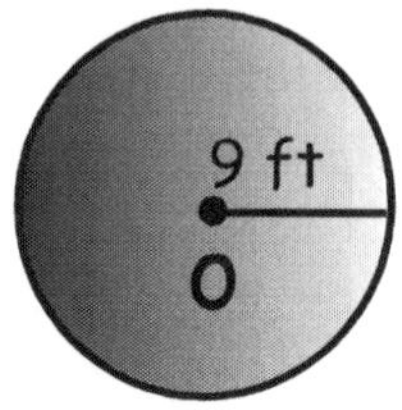

(A) 8147.1 ft^3 (B) 7084.4 ft^3

(C) 3053.6 ft^3 (D) 3542.2 ft^3

14. Find the volume of the sphere with a radius of 5.6 in and round the answer to the nearest tenth.

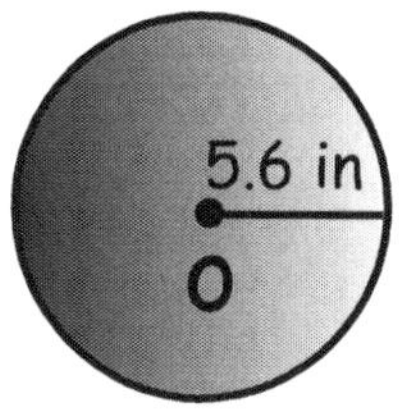

(A) 632.6 in^3 (B) 735.6 in^3

(C) 518.7 in^3 (D) 456.5 in^3

15. Find the volume of the sphere with a diameter of 26 ft and round the answer to the nearest tenth.

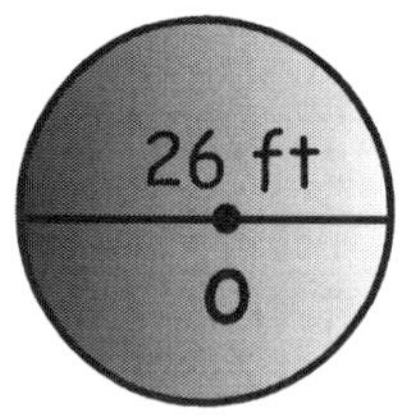

(A) 5383.6 ft^3 (B) 4737.6 ft^3

(C) 9202.8 ft^3 (D) 10767.2 ft^3

1. Find the volume of the rectangular prism measuring 1 ft and 4 ft along the base and 2 ft tall. Round the answer to the nearest tenth.

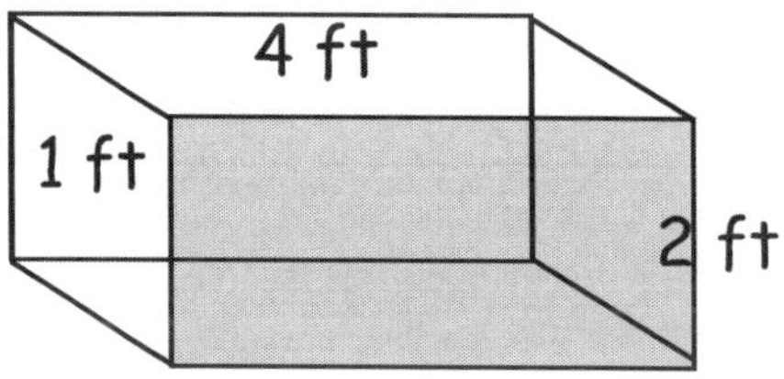

(A) 8 ft^3 (B) 4 ft^3

(C) 9.2 ft^3 (D) 4.6 ft^3

2. Find the volume of the rectangular prism measuring 15 in. and 18 in. along the base and 15 in. tall. Round the answer to the nearest tenth.

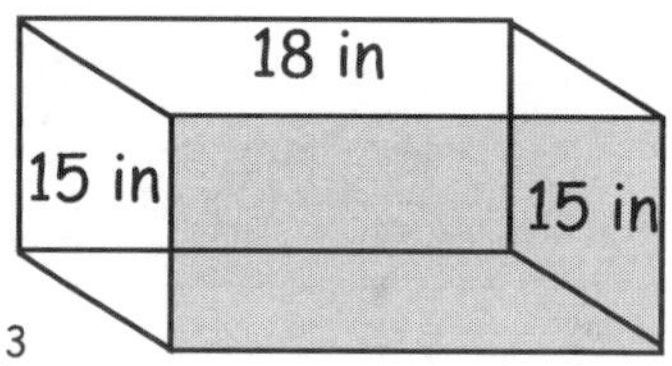

(A) 3361.5 in^3 (B) 2689.2 in^3

(C) 3173.3 in^3 (D) 4050 in^3

3. Find the volume of a square prism measuring 3 yd along each edge of the base and 17 yd tall. Round the answer to the nearest tenth.

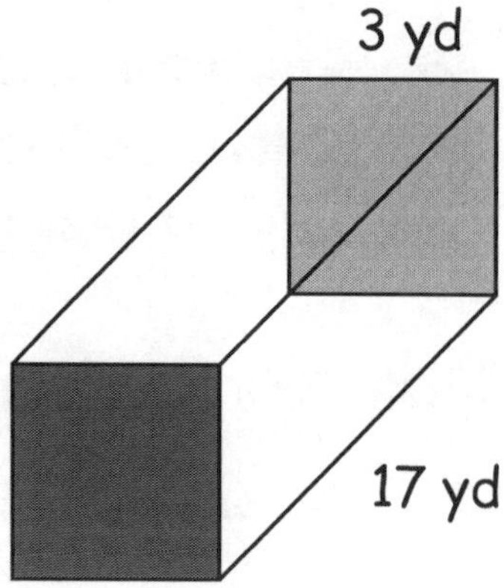

(A) 128.5 yd^3 (B) 153 yd^3

(C) 115.8 yd^3 (D) 131.6 yd^3

4. Find the volume of a rectangular prism measuring 11 ft and 3 ft along the base and 7 ft tall. Round the answer to the nearest tenth.

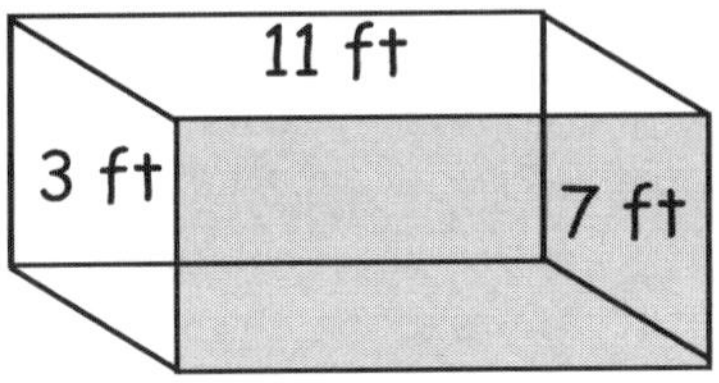

(A) 924 ft^3 (B) 462 ft^3

(C) 1848 ft^3 (D) 231 ft^3

5. Find the volume of a square prism measuring 17 km along each edge of the base and 19 km tall. Round the answer to the nearest tenth.

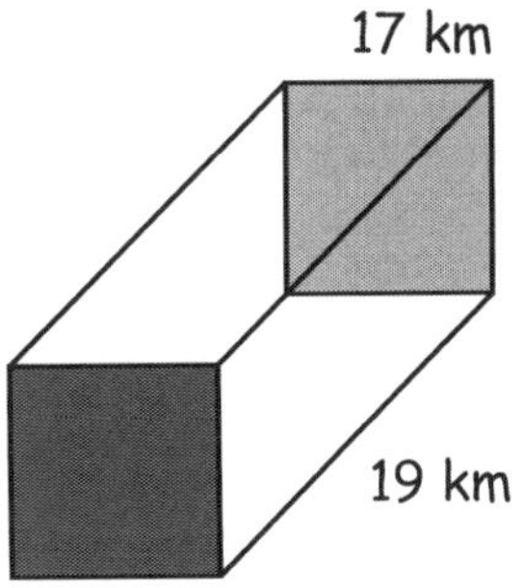

(A) 4101.8 km^3 (B) 5491 km^3

(C) 3363.5 km^3 (D) 4557.5 km^3

6. Find the volume of a square prism measuring 10 cm along each edge of the base and 7 cm tall. Round the answer to the nearest tenth.

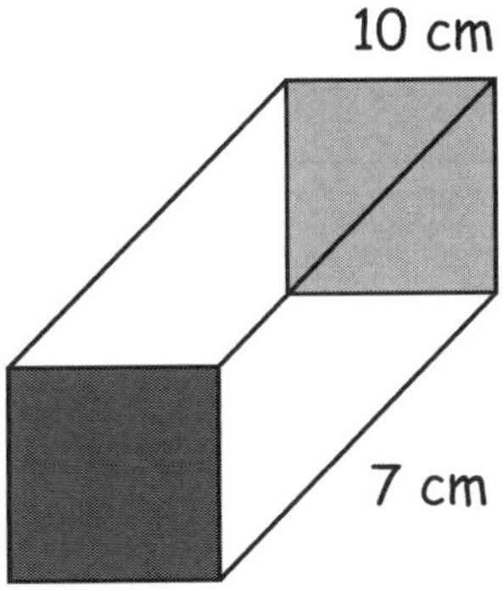

(A) 1416.2 cm^3 (B) 833 cm^3

(C) 708.1 cm^3 (D) 700 cm^3

7. Find the volume of a square prism measuring 12 in. along each edge of the base and 14 in. tall. Round the answer to the nearest tenth.

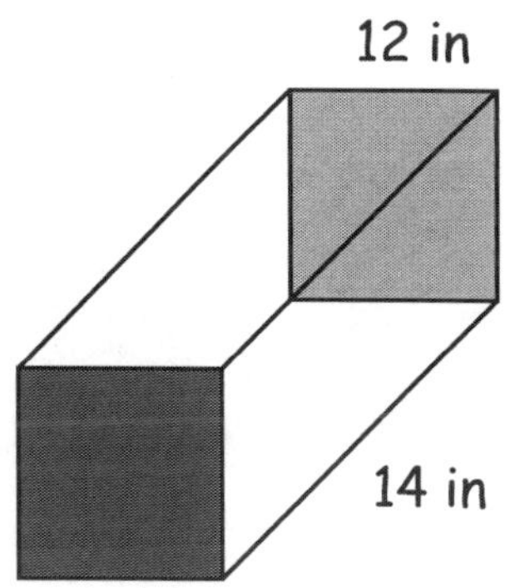

(A) 2016 in^3 (B) 4032 in^3

(C) 4636.8 in^3 (D) 2318.4 in^3

8. Find the volume of a rectangular prism measuring 18 m and 7 m along the base and 16 m tall. Round the answer to the nearest tenth.

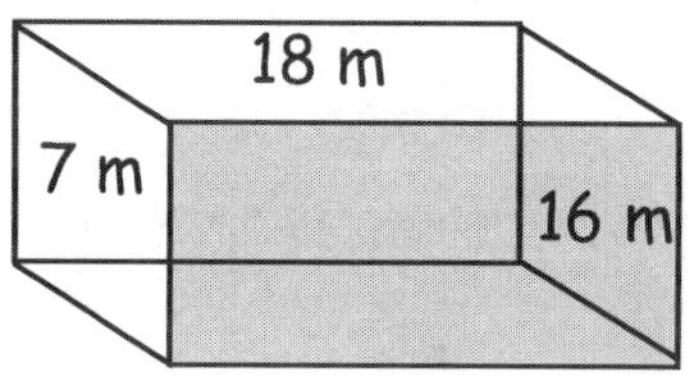

(A) 2016 m^3 (B) 4032 m^3

(C) 1633 m^3 (D) 1453.4 m^3

9. Find the volume of a rectangular prism measuring 17 cm and 10 cm along the base and 10 cm tall. Round the answer to the nearest tenth.

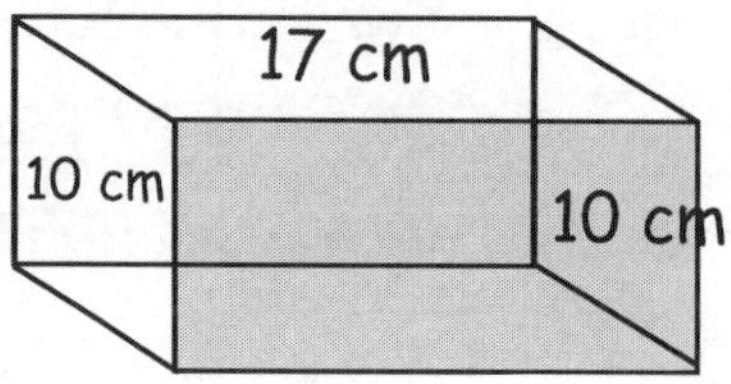

(A) 1445 cm^3 (B) 1700 cm^3

(C) 1891.6 cm^3 (D) 1719.6 cm^3

10. Find the volume of a square prism measuring 3 km along each edge of the base and 6 km tall. Round the answer to the nearest tenth.

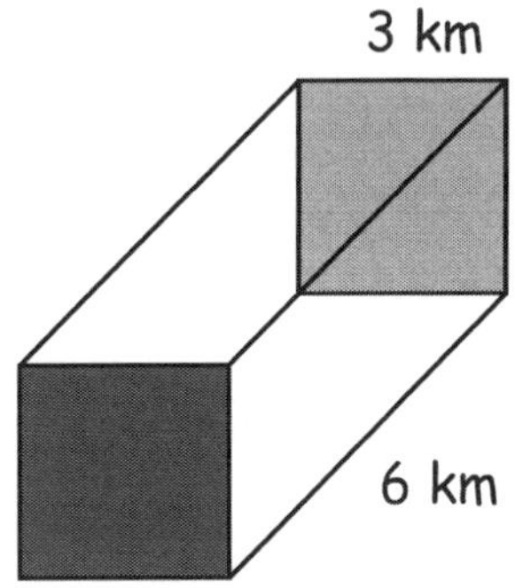

(A) 16 km^3 (B) 31.9 km^3

(C) 63.7 km^3 (D) 54 km^3

11. Find the volume of a square prism measuring 13 mm along each edge of the base and 6 mm tall. Round the answer to the nearest tenth.

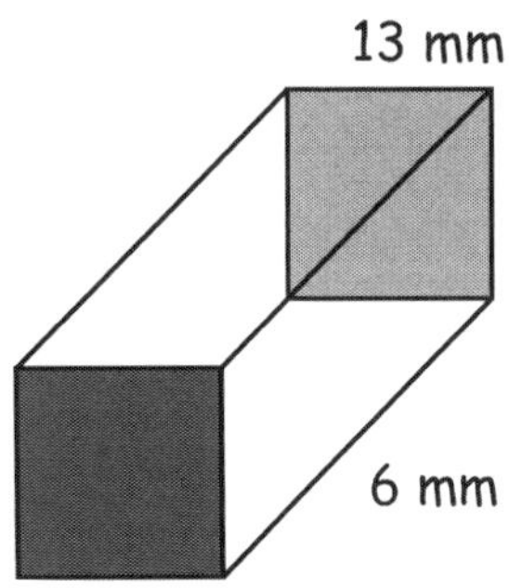

(A) 1216.8 mm^3 (B) 1058.6 mm^3

(C) 1014 mm^3 (D) 942.2 mm^3

12. Find the volume of a rectangular prism measuring 4 yd and 19 yd along the base and 5 yd tall. Round the answer to the nearest tenth.

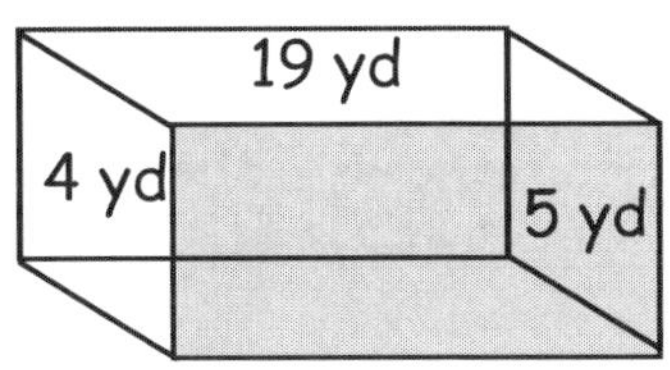

(A) 380 yd^3 (B) 334.4 yd^3

(C) 250.2 yd^3 (D) 294.3 yd^3

13. Find the volume of a rectangular prism measuring 9 cm and 5 cm along the base and 9cm in tall. Round the answer to the nearest tenth.

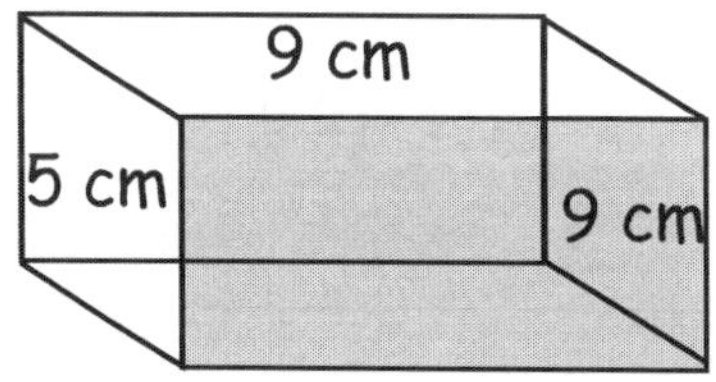

(A) 624 cm (B) $265.7\ cm^3$

(C) 405 cm (D) $236.5\ cm^3$

14. Find the volume of a rectangular prism measuring 20 mm and 10 mm along the base and 9 mm tall. Round the answer to the nearest tenth.

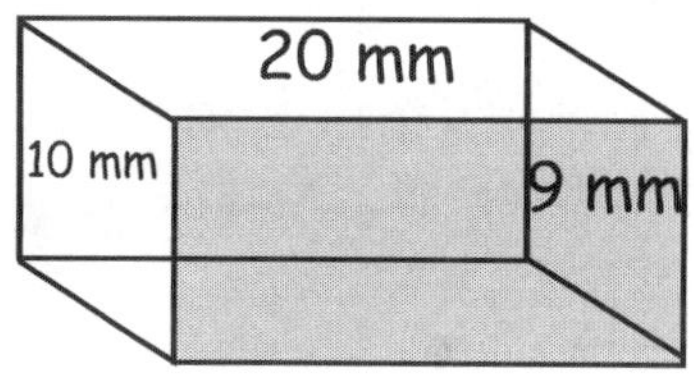

(A) $2880\ mm^3$ (B) $5760\ mm^3$

(C) $3600\ mm^3$ (D) $1800\ mm^3$

15. Find the volume of a square prism measuring 12 in along each edge of the base and 3 in tall. Round the answer to the nearest tenth.

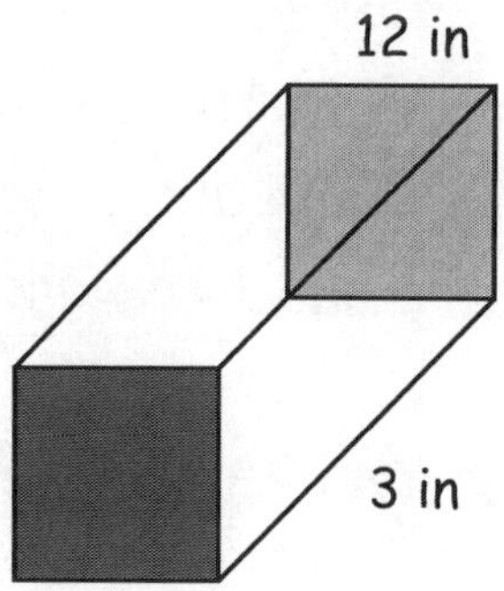

(A) $488.2\ in^3$ (B) $466\ in^3$

(C) $541.9\ in^3$ (D) $432\ in^3$

1. Find the volume of a cylinder with a diameter of 34 ft and a height of 12 ft. Round the answer to the nearest tenth.

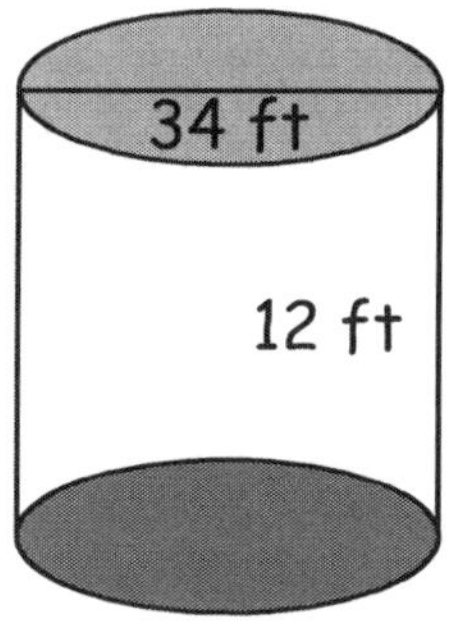

(A) 21598.4 ft^3 (B) 10895 ft^3

(C) 21790.1 ft^3 (D) 18303.7 ft^3

2. Find the volume of a cylinder with a radius of 6 mm and a height of 6 mm. Round the answer to the nearest tenth.

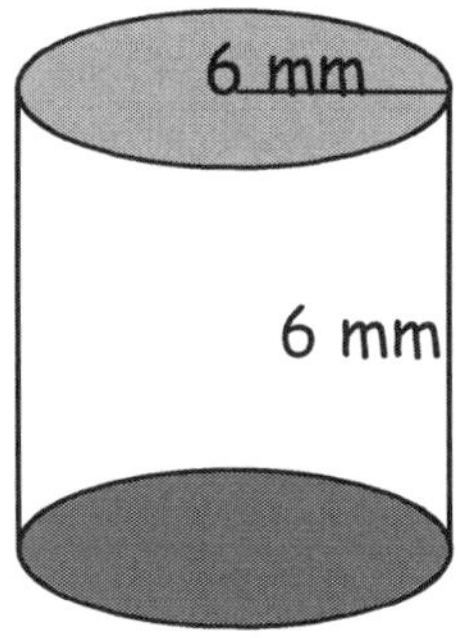

(A) 678.6 mm^3 (B) 3013 mm^3

(C) 1357.2 mm^3 (D) 2714.4 mm^3

3. Find the volume of a cylinder with a diameter of 22 in and a height of 15 in. Round the answer to the nearest tenth.

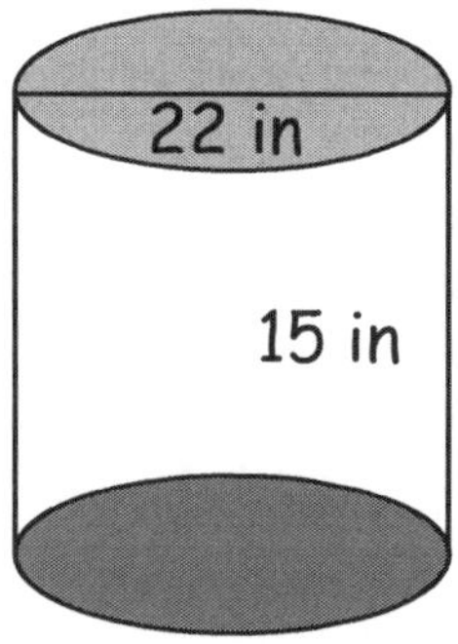

(A) 2394.9 in^3 (B) 4789.7 in^3

(C) 5702 in^3 (D) 1197.5 in^3

4. Find the volume of a cone with a diameter of 8 m and a height of 10 m. Round the answer to the nearest tenth.

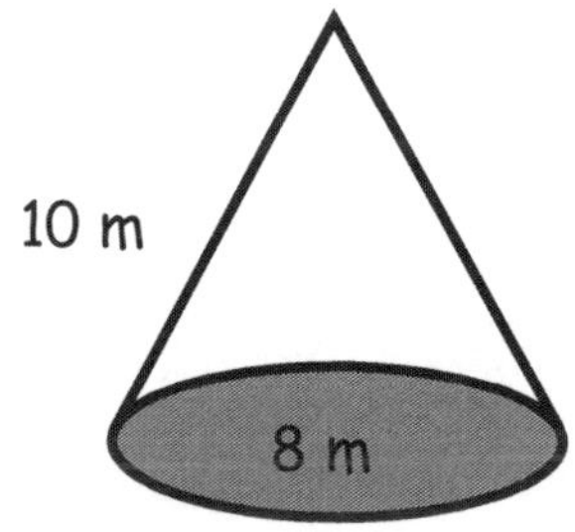

(A) 335.1 m^3

(B) 402.1 m^3

(C) 167.6 m^3

(D) 804.2 m^3

5. Find the volume of a cone with a radius of 3 km and a height of 10 km. Round the answer to the nearest tenth.

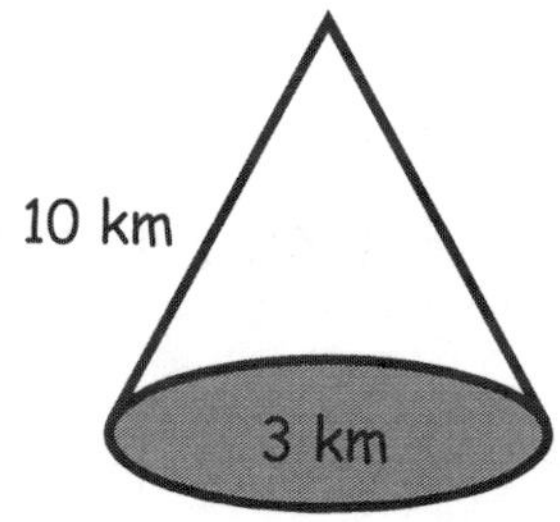

(A) 108.4 km^3

(B) 94.2 km^3

(C) 98.9 km^3

(D) 119.2 km^3

6. Find the volume of a cylinder with a radius of 14 yd and a height of 9 yd. Round the answer to the nearest tenth.

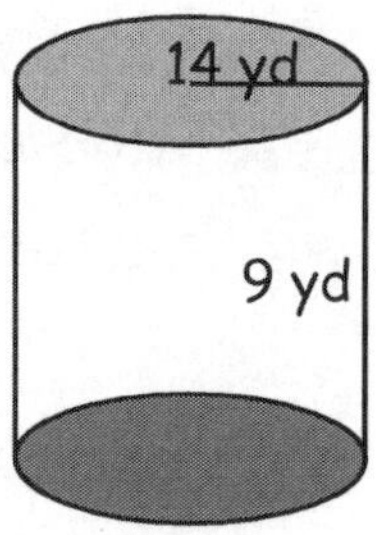

(A) 5541.8 yd^3

(B) 4876.8 yd^3

(C) 2770.9 yd^3

(D) 2438.4 yd^3

7. Find the volume of a cone with a diameter of 4 in and a height of 15 in. Round the answer to the nearest tenth.

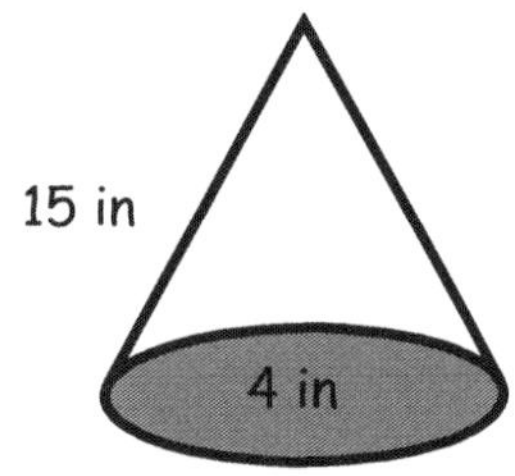

(A) 43.8 in^3 (B) 55.9 in^3

(C) 62.8 in^3 (D) 49.8 in^3

8. Find the volume of a cylinder with a diameter of 16 m and a height of 7 m. Round the answer to the nearest tenth.

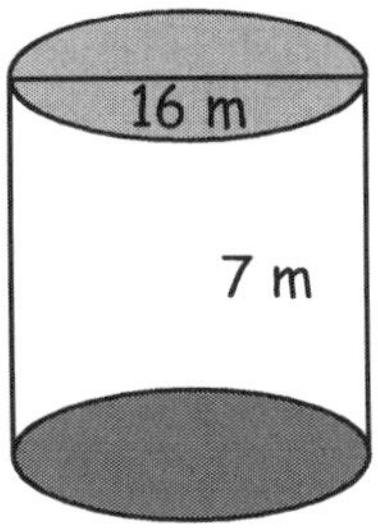

(A) 1407.4 m^3 (B) 757.2 m^3

(C) 934.8 m^3 (D) 1140 m^3

9. Find the volume of a cylinder with a radius of 9 ft and a height of 13 ft. Round the answer to the nearest tenth.

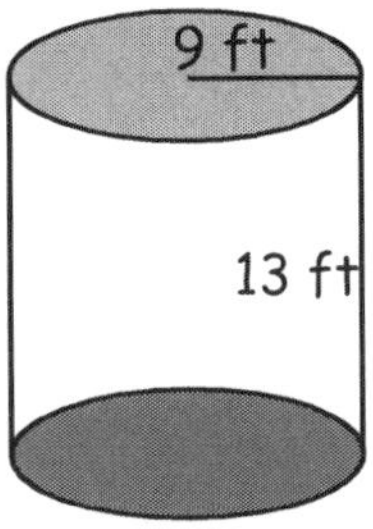

(A) 3308.1 ft^3 (B) 3804.3 ft^3

(C) 6695.6 ft^3 (D) 3347.8 ft^3

10. Find the volume of a cylinder with a diameter of 10 yd and a height of 3 yd. Round the answer to the nearest tenth.

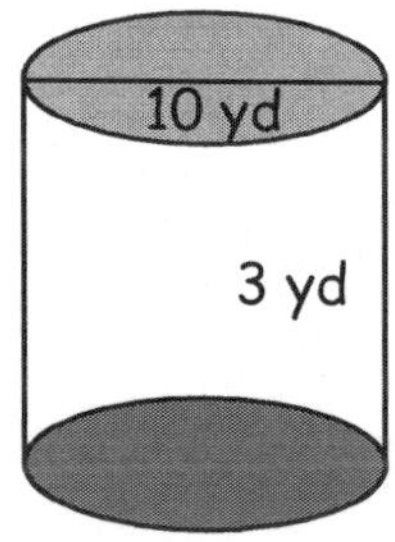

(A) 228 yd^3 (B) 235.6 yd^3

(C) 193.2 yd^3 (D) 193.8 yd^3

11. Find the volume of a cylinder with a diameter of 40 ft and a height of 11 ft. Round the answer to the nearest tenth.

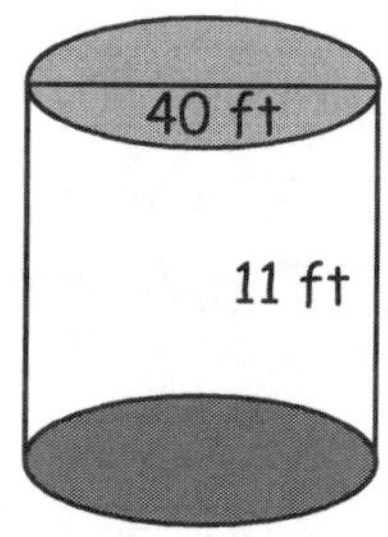

(A) 11749.6 ft^3 (B) 9869.7 ft^3

(C) 13823 ft^3 (D) 4934.9 ft^3

12. Find the volume of a cone with a diameter of 6 km and a height of 11 km. Round the answer to the nearest tenth.

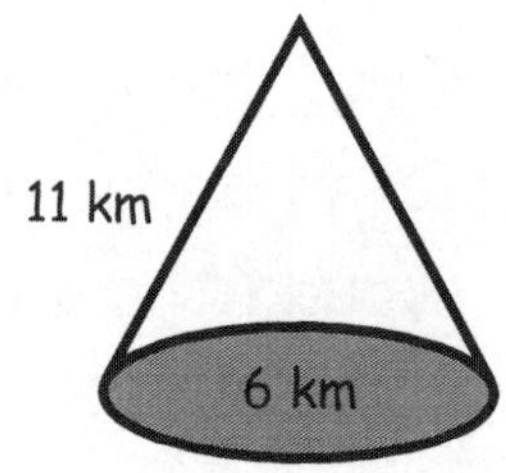

(A) 103.7 km^3 (B) 51.8 km^3

(C) 25.9 km^3 (D) 21.8 km^3

13. Find the volume of a cone with a diameter of 6 m and a height of 8 m. Round the answer to the nearest tenth.

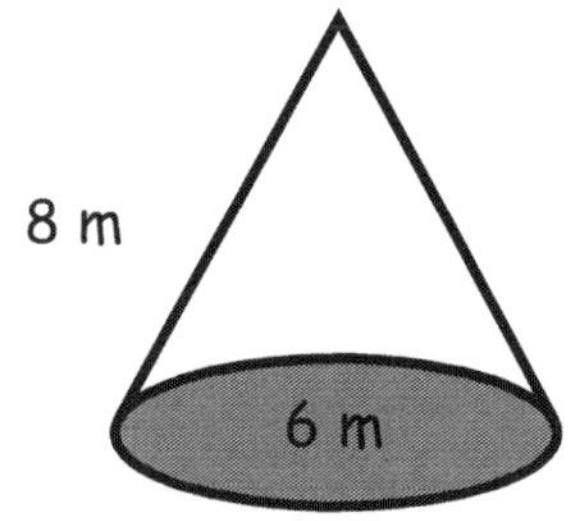

(A) 65.6 m^3 (B) 91.4 m^3

(C) 75.4 m^3 (D) 78.1 m^3

14. Find the volume of a cylinder with a radius of 20 in and a height of 16 in. Round the answer to the nearest tenth.

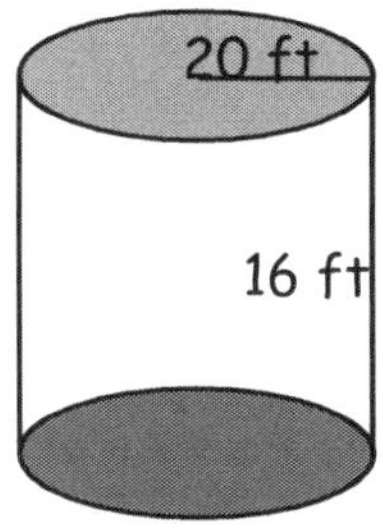

(A) 20166.5 in^3 (B) 20106.2 in^3

(C) 23725.3 in^3 (D) 24199.8 in^3

15. Find the volume of a cylinder with a diameter of 16 ft and a height of 8 ft. Round the answer to the nearest tenth.

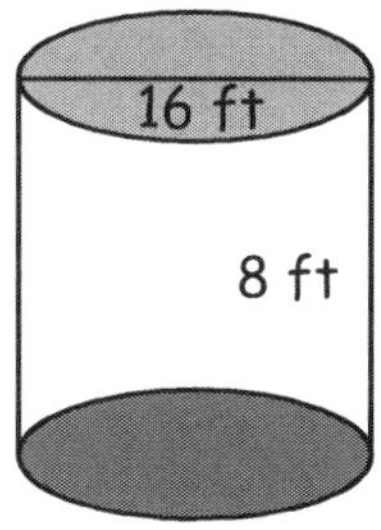

(A) 1865.9 ft^3 (B) 466.5 ft^3

(C) 1608.5 ft^3 (D) 933 ft^3

Find the additive inverse of the below questions.

1. 8

(A) $\frac{1}{8}$ (B) $-\frac{1}{8}$ (C) 0 (D) -8

2. -11

(A) 11 (B) 0 (C) $-\frac{1}{11}$ (D) $\frac{1}{11}$

3. -532

(A) 0 (B) -532 (C) 532 (D) $\frac{1}{532}$

4. 6921

(A) $\frac{1}{6921}$ (B) $-\frac{1}{6921}$ (C) -6921 (D) 0

Find the additive inverse of the below questions.

5. 7751

(A) $\frac{1}{7751}$ (B) -7751 (C) 0 (D) $-\frac{1}{7751}$

6. -9021

(A) -9021 (B) 0 (C) $-\frac{1}{9021}$ (D) 9021

7. -1431

(A) 1431 (B) $-\frac{1}{1431}$ (C) 0 (D) $\frac{1}{1431}$

8. $\frac{1}{2}$

(A) $-\frac{1}{2}$ (B) 0 (C) 2 (D) -2

Find the additive inverse of the below questions.

9. $-\frac{2}{3}$

(A) 0 (B) $\frac{3}{2}$ (C) $\frac{2}{3}$ (D) $-\frac{3}{2}$

10. $\frac{5}{6}$

(A) $\frac{6}{5}$ (B) $-\frac{5}{6}$ (C) 0 (D) $-\frac{6}{5}$

11. $\frac{7}{17}$

(A) $-\frac{7}{17}$ (B) $\frac{17}{7}$ (C) $-\frac{17}{7}$ (D) 0

12. $-\frac{4}{5}$

(A) 0 (B) $\frac{5}{4}$ (C) $-\frac{5}{4}$ (D) $\frac{4}{5}$

Find the additive inverse of the below questions.

13. $\frac{9}{19}$

(A) $-\frac{19}{9}$ (B) $\frac{19}{9}$ (C) $-\frac{9}{19}$ (D) 0

14. $-\frac{3}{7}$

(A) 0 (B) $\frac{3}{7}$ (C) $\frac{7}{3}$ (D) $-\frac{7}{3}$

15. 4521

(A) 4521 (B) 0 (C) $\frac{1}{4521}$ (D) -4521

Find the multiplicative inverse of the below questions.

1. 5

(A) $\frac{1}{5}$ (B) $-\frac{1}{5}$ (C) 0 (D) -5

2. -11

(A) 11 (B) 0 (C) $-\frac{1}{11}$ (D) $\frac{1}{11}$

3. -532

(A) 0 (B) 532 (C) $-\frac{1}{532}$ (D) $\frac{1}{532}$

4. 903

(A) $-\frac{1}{903}$ (B) $\frac{1}{903}$ (C) -903 (D) 0

Find the multiplicative inverse of the below questions.

5. 620

(A) $\frac{1}{620}$ (B) $-\frac{1}{620}$ (C) 0 (D) -620

6. 2

(A) 0 (B) -2 (C) $-\frac{1}{2}$ (D) $\frac{1}{2}$

7. 17

(A) $-\frac{1}{17}$ (B) $\frac{1}{17}$ (C) 0 (D) -17

8. $\frac{1}{2}$

(A) $-\frac{1}{2}$ (B) 0 (C) 2 (D) -2

Find the multiplicative inverse of the below questions.

9. $-\frac{3}{7}$

(A) $-2\frac{1}{3}$ (B) $\frac{3}{7}$ (C) $\frac{7}{3}$ (D) 0

10. $\frac{1}{11}$

(A) -11 (B) $-\frac{1}{11}$ (C) 0 (D) 11

11. -303

(A) $\frac{1}{303}$ (B) 303 (C) $-\frac{1}{303}$ (D) 0

12. $-\frac{1}{9}$

(A) 0 (B) 9 (C) -9 (D) $\frac{1}{9}$

Find the multiplicative inverse of the below questions.

13. $\frac{21}{47}$

(A) $-2\frac{5}{21}$ (B) $\frac{47}{21}$ (C) $-\frac{47}{19}$ (D) 0

14. $-\frac{5}{29}$

(A) 0 (B) $\frac{29}{5}$ (C) $\frac{29}{5}$ (D) $-5\frac{4}{5}$

15. 57

(A) 57 (B) 0 (C) $\frac{1}{57}$ (D) -57

1. Which of the below is a rational number ?

(A) $\sqrt{9}$ (B) $\sqrt{5}$ (C) $\sqrt{11}$ (D) $\sqrt{6}$

2. Which of the below is a rational number ?

(A) $\sqrt{80}$ (B) $\sqrt{42}$ (C) $\sqrt{35}$ (D) $\sqrt{81}$

3. Which of the below is a rational number ?

(A) $\sqrt{25}$ (B) $\sqrt{21}$ (C) $\sqrt{30}$ (D) $\sqrt{19}$

4. Which of the below is a rational number ?

(A) $\sqrt{3}$ (B) $\sqrt{111}$ (C) $\sqrt{4}$ (D) $\sqrt{216}$

5. Which of the below is a rational number ?

(A) $\sqrt{32}$ (B) $\sqrt{49}$ (C) $\sqrt{41}$ (D) $\sqrt{99}$

6. Which of the below is a rational number ?

(A) $\sqrt{21}$ (B) $\sqrt{63}$ (C) $\sqrt{88}$ (D) $\sqrt{49}$

7. Which of the below is an irrational number ?

(A) $\sqrt{121}$ (B) $\sqrt{117}$ (C) $\sqrt{169}$ (D) $\sqrt{144}$

8. Which of the below is an irrational number ?

(A) $\sqrt{81}$ (B) $\sqrt{64}$ (C) $\sqrt{400}$ (D) $\sqrt{5}$

9. Which of the below is an irrational number ?

(A) $\sqrt{7}$ (B) $\sqrt{16}$ (C) $\sqrt{49}$ (D) $\sqrt{900}$

10. Which of the below is an irrational number ?

(A) $\sqrt{225}$ (B) $\sqrt{25}$ (C) $\sqrt{23}$ (D) $\sqrt{36}$

11. Which of the below is an irrational number ?

(A) $\sqrt{256}$ (B) $\sqrt{39}$ (C) $\sqrt{16}$ (D) $\sqrt{289}$

12. Which of the below is an irrational number ?

(A) $\sqrt{441}$ (B) $\sqrt{686}$ (C) $\sqrt{484}$ (D) $\sqrt{84}$

13. Which of the below is an irrational number ?

(A) $\sqrt{71}$ (B) $\sqrt{1089}$ (C) $\sqrt{961}$ (D) $\sqrt{196}$

14. Which of the below is an irrational number ?

(A) $\sqrt{2500}$ (B) $\sqrt{1600}$ (C) $\sqrt{11}$ (D) $\sqrt{121}$

15. Which of the below is an irrational number ?

(A) $\sqrt{4}$ (B) $\sqrt{45}$ (C) $\sqrt{16}$ (D) $\sqrt{100}$

Simplify the below. Your answer should contain only positive exponents.

1. $4x^2 \cdot 3x^0$

(A) $12x^3$ (B) $9x^3$ (C) $12x^2$ (D) $3x^3$

2. $3nn^2$

(A) $3n^3$ (B) $2n^3$ (C) $12n^3$ (D) $8n$

3. $2r^0 \cdot 3r^2$

(A) $6r^2$ (B) $12r^6$ (C) $12r^2$ (D) $48r^5$

4. $3p^2 \cdot 2p$

(A) $12p^3$ (B) $9p^3$ (C) $12p^2$ (D) $6p^3$

Simplify the below. Your answer should contain only positive exponents.

5. $x \cdot 4x$

(A) $12x^4$ (B) $4x^2$ (C) $6x^3$ (D) $12x^3$

6. $3m^2 \cdot 4m$

(A) $12m^3$ (B) $16m^3$ (C) $2m^3$ (D) $2m^4$

7. $2n^2 \cdot 2n^2$

(A) $3n^2$ (B) $6n^4$ (C) $4n^4$ (D) $4n^3$

8. $2x^4 \cdot 4x^2$

(A) $2x^3$ (B) $6x^3$ (C) $8x^6$ (D) $4x^3$

Simplify the below. Your answer should contain only positive exponents.

9. $4v^2 \cdot 2v^2$

(A) $3v^3$ (B) $8v^2$ (C) $16v^4$ (D) $8v^4$

10. $3n \cdot 4n^2$

(A) $9n$ (B) $8n^3$ (C) $8n^4$ (D) $12n^3$

11. $n \cdot 4n \cdot 3n$

(A) $12n^3$ (B) n^4 (C) $2n^4$ (D) $6n^2$

12. $2v^2 \cdot 2v^5$

(A) $3v^2$ (B) $2v^3$ (C) $24v^5$ (D) $4v^7$

Simplify the below. Your answer should contain only positive exponents.

13. $3b^2 \cdot 4b^0$

(A) $8b^3$ (B) $4b$ (C) $12b^2$ (D) $8b^4$

14. $2x \cdot 3x$

(A) $8x^3$ (B) $6x^2$ (C) $9x^4$ (D) $12x^2$

15. $x^2 \cdot x$

(A) $12x^2$ (B) x^3 (C) $12x^3$ (D) $24x^5$

Find the slope of the straight line passing through the points given below.

1. A(18, −15), B(18, −17)

(A) Undefined

(B) −1

(C) 1

(D) 0

2. M(−9, − 14), N(−9, 7)

(A) $-\frac{5}{2}$

(B) Undefined

(C) 0

(D) $\frac{5}{2}$

3. O(−1, 1), P(−15, −1)

(A) $-\frac{1}{7}$

(B) $\frac{1}{7}$

(C) −7

(D) 7

Find the slope of the straight line passing through the points given below.

4. Q(-1, 9), R(9, -6)

(A) $\frac{2}{3}$

(B) $-\frac{2}{3}$

(C) $-\frac{3}{2}$

(D) $\frac{3}{2}$

5. S(-17, 20), T(-16, -17)

(A) -37

(B) $-\frac{1}{37}$

(C) $\frac{1}{37}$

(D) 37

6. U(-6, 12), V(-6, 8)

(A) -1

(B) 1

(C) Undefined

(D) 0

Find the slope of the straight line passing through the points given below.

7. X(−20, 2), Y(−20, 1)

(A) 0

(B) −2

(C) 2

(D) Undefined

8. C(−13, 16), D(8, 1)

(A) $\frac{7}{5}$

(B) $-\frac{7}{5}$

(C) $\frac{5}{7}$

(D) $-\frac{5}{7}$

9. E(15, 8), F(17, −19)

(A) $\frac{27}{2}$

(B) $-\frac{2}{27}$

(C) $\frac{2}{27}$

(D) $-\frac{27}{2}$

Find the slope of the straight line passing through the points given below.

10. H(−11, 8), I(14, −2)

(A) $\frac{5}{2}$

(B) $-\frac{2}{5}$

(C) $-\frac{5}{2}$

(D) $\frac{2}{5}$

11. J(−7, −14), K(−13, 13)

(A) $-\frac{2}{9}$

(B) $\frac{2}{9}$

(C) $-\frac{9}{2}$

(D) $\frac{9}{2}$

12. L(5, 14), M(−17, −17)

(A) $-\frac{22}{31}$

(B) $-\frac{31}{22}$

(C) $\frac{31}{22}$

(D) $\frac{22}{31}$

Find the slope of the straight line passing through the points given below.

13. R(1, 13), S(−9, −4)

(A) $\frac{17}{10}$

(B) $\frac{10}{17}$

(C) $-\frac{10}{17}$

(D) $-\frac{17}{10}$

14. N(−6, −15), M(−16, −15)

(A) $-\frac{3}{4}$

(B) $\frac{3}{4}$

(C) 0

(D) Undefined

15. I(15, −15), J(−5, −7)

(A) $-\frac{2}{5}$

(B) $-\frac{5}{2}$

(C) $\frac{2}{5}$

(D) $\frac{5}{2}$

Find the slope-intercept form of the straight line with the given slope and y-intercept for all the problems given below.

1. Slope = $-\frac{2}{5}$, y-intercept = 5

(A) $y = -5x - \frac{2}{5}$

(B) $y = -\frac{2}{5}x + 5$

(C) $y = 5x - \frac{2}{5}$

(D) $y = -\frac{2}{5}x - 5$

2. Slope = $\frac{7}{4}$, y-intercept = 5

(A) $y = x + 5$

(B) $y = 5x + 1$

(C) $y = \frac{7}{4}x + 5$

(D) $y = -x + 1$

3. Slope = $-\frac{9}{4}$, y-intercept = -4

(A) $y = -4x + \frac{1}{2}$

(B) $y = -\frac{9}{4}x - 4$

(C) $y = -\frac{1}{2}x - 4$

(D) $y = \frac{1}{2}x - 4$

Find the slope-intercept form of the straight line with the given slope and y-intercept for all the problems given below.

4. Slope = –3, y-intercept = 5

(A) $y = -3x + 5$

(B) $y = 5x - 3$

(C) $y = -5x - 3$

(D) $y = -3x - 5$

5. Slope = 6, y-intercept = 3

(A) $y = -5x - 6$

(B) $y = 3x - 6$

(C) $y = -6x + 3$

(D) $y = 6x + 3$

6. Slope = $\frac{3}{5}$, y-intercept = –4

(A) $y = \frac{3}{5}x + 4$

(B) $y = 4x + \frac{3}{5}$

(C) $y = \frac{3}{5}x - 4$

(D) $y = -4x + \frac{3}{5}$

Find the slope-intercept form of the straight line with the given slope and y-intercept for all the problems given below.

7. Slope = $\frac{-1}{2}$, y-intercept = -3

(A) $y = -3x - \frac{1}{2}$

(B) $y = -2x - 3$

(C) $y = -\frac{1}{2}x - 3$

(D) $y = \frac{2}{5}x - 3$

8. Slope = -1, y-intercept = -1

(A) $y = 3x - 1$

(B) $y = -3x - 1$

(C) $y = -x - 3$

(D) $y = -x - 1$

9. Slope = $\frac{4}{3}$, y-intercept = 1

(A) $y = \frac{4}{3}x + 1$

(B) $y = \frac{4}{3}x - 1$

(C) $y = x + \frac{4}{3}$

(D) $y = -x + \frac{4}{3}$

Find the slope-intercept form of the straight line with the given slope and y-intercept for all the problems given below.

10. Slope = $-\frac{6}{5}$, y-intercept = -4

(A) $y = -4x - \frac{2}{5}$

(B) $y = -\frac{2}{5}x - 4$

(C) $y = \frac{4}{5}x - 4$

(D) $y = -\frac{6}{5}x - 4$

11. Slope = $-\frac{7}{4}$, y-intercept = -3

(A) $y = \frac{7}{4}x - 3$

(B) $y = -\frac{7}{4}x - 3$

(C) $y = 3x + \frac{7}{4}$

(D) $y = -3x + \frac{7}{4}$

12. Slope = $-\frac{7}{5}$, y-intercept = 4

(A) $y = -\frac{7}{5}x + 4$

(B) $y = 4x + \frac{7}{5}$

(C) $y = -4x + \frac{7}{5}$

(D) $y = \frac{7}{5}x + 4$

Find the slope-intercept form of the straight line with the given slope and y-intercept for all the problems given below.

13. Slope = $-\frac{2}{3}$, y-intercept = 2

(A) $y = -\frac{2}{3}x + 2$

(B) $y = 2x - \frac{1}{3}$

(C) $y = \frac{1}{3}x + 2$

(D) $y = -\frac{1}{3}x + 2$

14. Slope = $\frac{9}{2}$, y-intercept = –5

(A) $y = \frac{9}{2}x - 5$

(B) $y = -4x - 5$

(C) $y = \frac{1}{2}x - 5$

(D) $y = -2x - 5$

15. Slope = 1, y-intercept = –1

(A) $y = x - 1$

(B) $y = -x - 1$

(C) $y = 3x - 1$

(D) $y = -3x - 1$

Find the slope of the straight line for each of the questions below.

1)

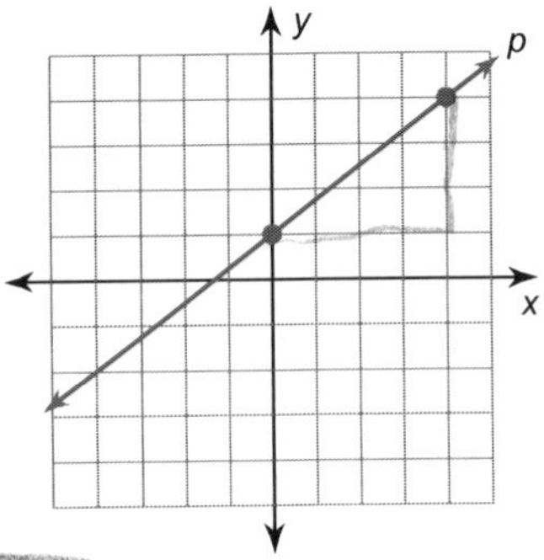

A) $\frac{4}{3}$ B) $-\frac{3}{4}$

C) $-\frac{4}{3}$ D) $\frac{3}{4}$

2)

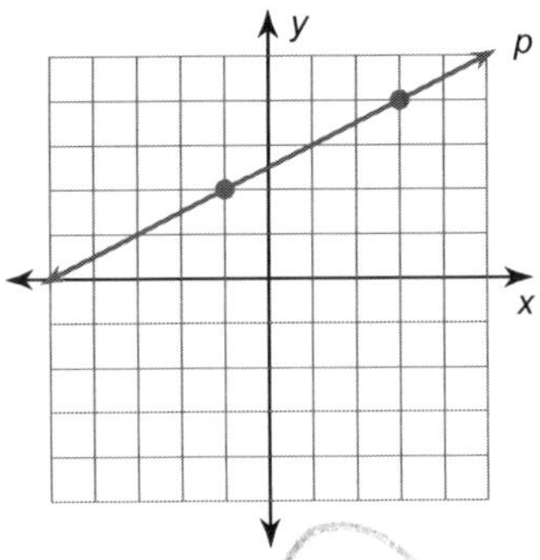

A) 2 B) $\frac{1}{2}$

C) −2 D) $-\frac{1}{2}$

3)

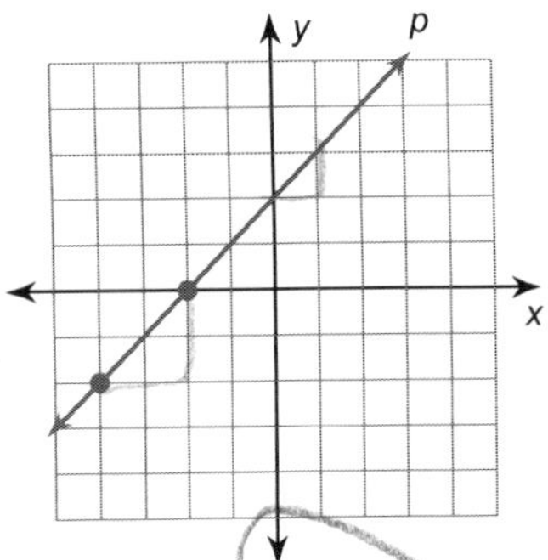

A) $-\frac{4}{5}$ B) 1

C) −1 D) $\frac{4}{5}$

4)

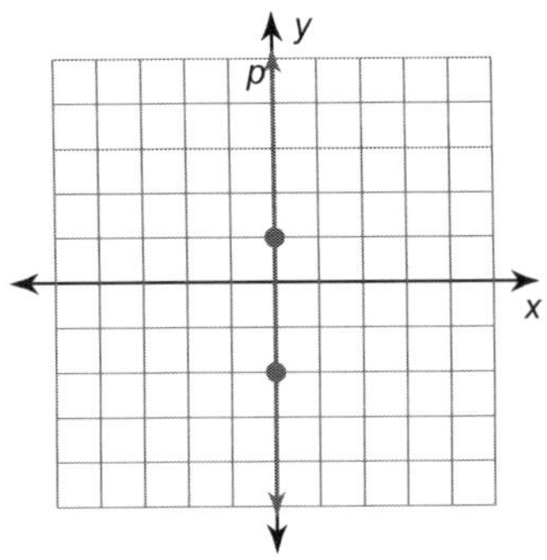

A) $\frac{3}{4}$ B) $-\frac{3}{4}$

C) 0 D) Undefined

Find the slope of the straight line for each of the questions below.

5)

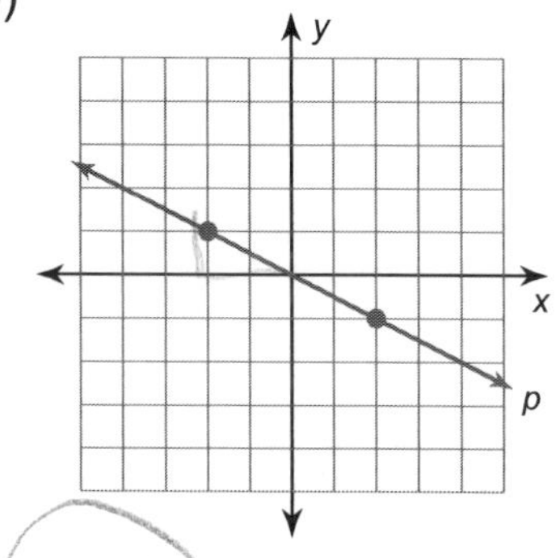

A) $-\frac{1}{2}$ B) $\frac{1}{2}$

C) -2 D) 2

6)

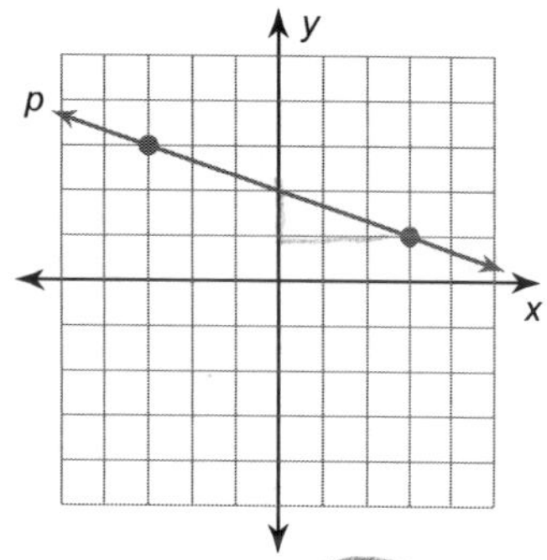

A) 3 B) $\frac{1}{3}$

C) -3 D) $-\frac{1}{3}$

7)

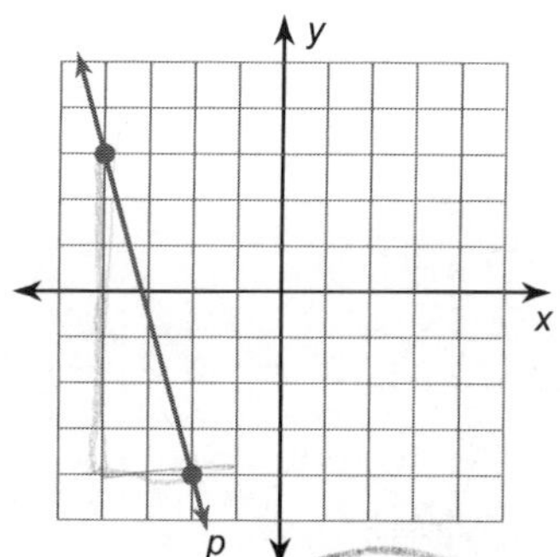

A) $\frac{7}{2}$ B) $-\frac{7}{2}$

C) $-\frac{2}{7}$ D) $\frac{2}{7}$

8)

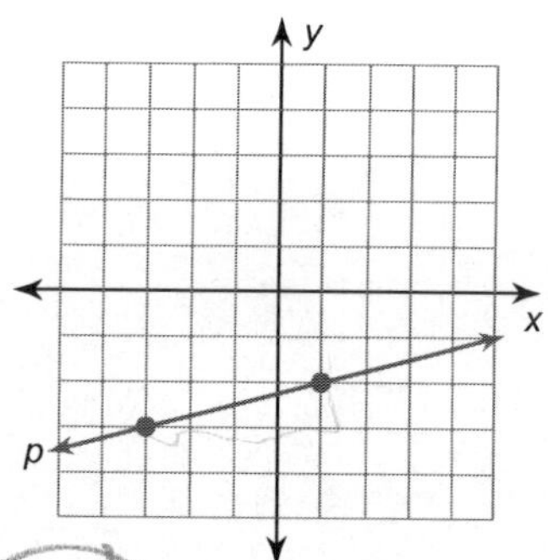

A) $\frac{1}{4}$ B) -4

C) 4 D) $-\frac{1}{4}$

Find the slope of the straight line for each of the questions below.

9)

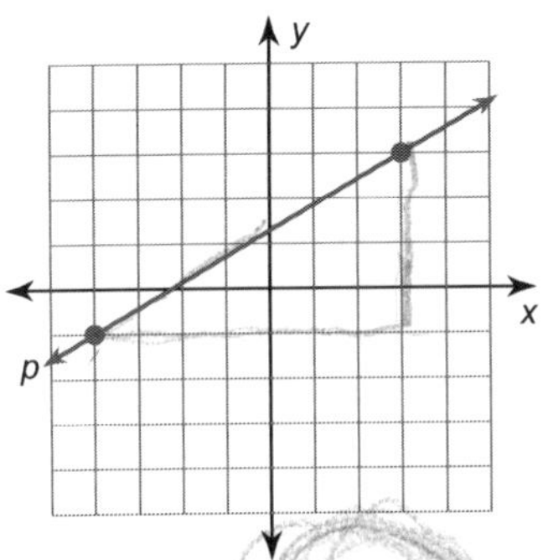

A) $\frac{7}{4}$ B) $\frac{4}{7}$

C) $-\frac{7}{4}$ D) $-\frac{4}{7}$

10)

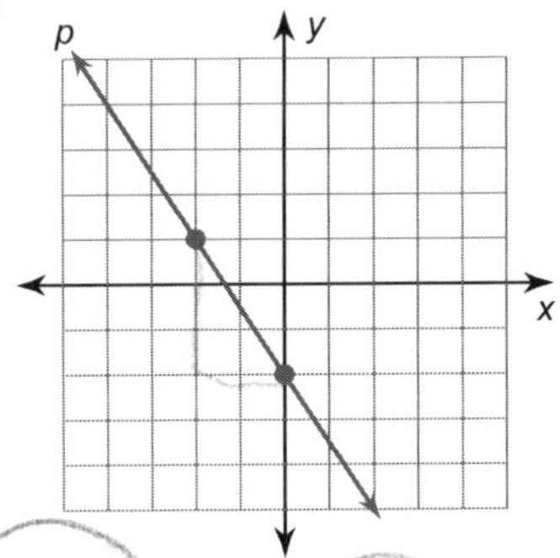

A) $-\frac{3}{2}$ B) $\frac{3}{2}$

C) $\frac{2}{3}$ D) $-\frac{2}{3}$

11)

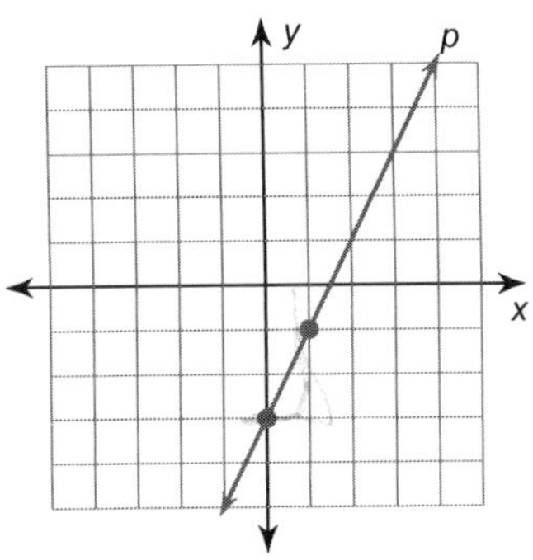

A) 2 B) $-\frac{1}{2}$

C) $\frac{1}{2}$ D) −2

12)

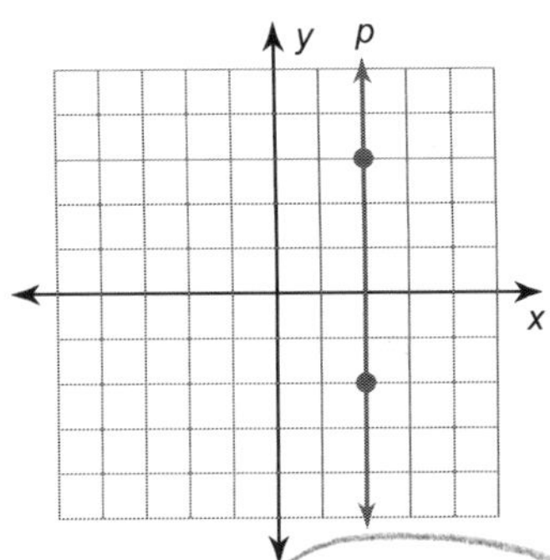

A) $-\frac{2}{5}$ B) Undefined

C) $\frac{2}{5}$ D) 0

Find the slope of the straight line for each of the questions below.

13)

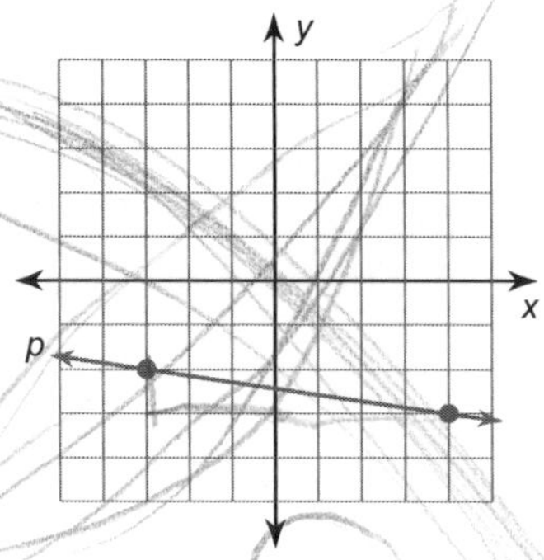

A) $\frac{1}{7}$ B) $-\frac{1}{7}$

C) 7 D) −7

14)

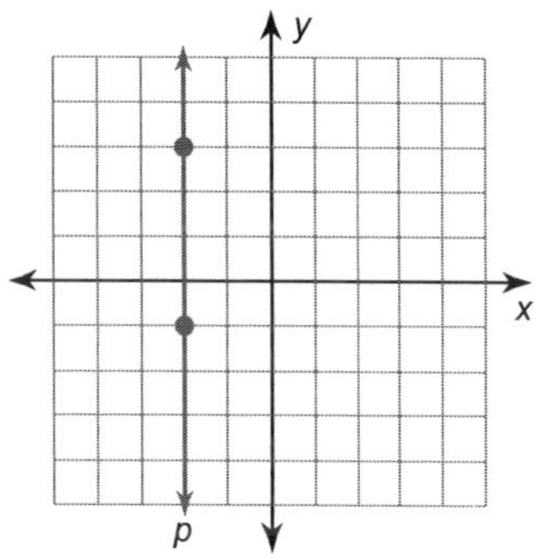

A) 0 B) $-\frac{1}{5}$

C) $\frac{1}{5}$ D) Undefined

15)

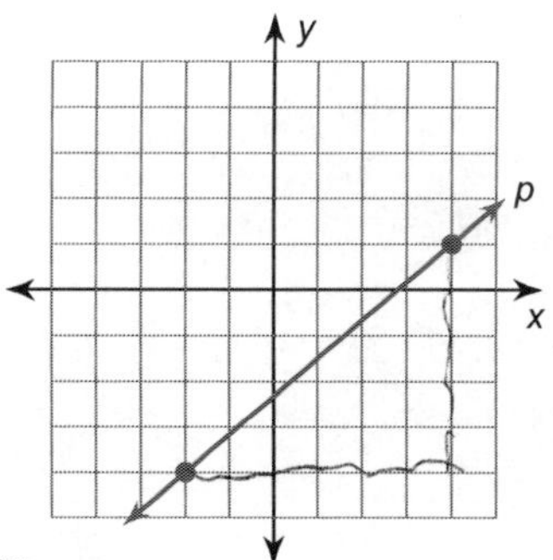

A) $\frac{5}{6}$ B) $\frac{6}{5}$

C) $-\frac{6}{5}$ D) $-\frac{5}{6}$

Find the slope of the straight line for each of the questions below.

1) $x + y = -3$

A) $-\frac{3}{2}$ B) 1

C) $\frac{3}{2}$ D) -1

2) $4x + 3y = 3$

A) $-\frac{4}{3}$ B) $-\frac{3}{4}$

C) $\frac{3}{4}$ D) $\frac{4}{3}$

3) $2x - 3y = 6$

A) $\frac{2}{3}$ B) $-\frac{2}{3}$

C) $\frac{3}{2}$ D) $-\frac{3}{2}$

4) $3x - 5y = -10$

A) $\frac{5}{3}$ B) $-\frac{5}{3}$

C) $-\frac{3}{5}$ D) $\frac{3}{5}$

5) $5x + 2y = 0$

A) $-\frac{5}{2}$ B) $-\frac{2}{5}$

C) $\frac{5}{2}$ D) $\frac{2}{5}$

6) $4x + y = 1$

A) 4 B) $\frac{1}{4}$

C) $-\frac{1}{4}$ D) -4

7) $x + 2y = 6$

A) 2 B) $\frac{1}{2}$

C) $-\frac{1}{2}$ D) -2

8) $5x - 3y = -12$

A) $-\frac{5}{3}$ B) $\frac{5}{3}$

C) $\frac{3}{5}$ D) $-\frac{3}{5}$

9) $7x-2y=8$

A) $-\frac{7}{2}$ B) $-\frac{2}{7}$

C) $\frac{7}{2}$ D) $\frac{2}{7}$

10) $5x-3y=0$

A) $\frac{5}{3}$ B) $-\frac{5}{3}$

C) $\frac{3}{5}$ D) $-\frac{3}{5}$

11) $3x-2y=-6$

A) $-\frac{2}{3}$ B) $\frac{3}{2}$

C) $\frac{2}{3}$ D) $-\frac{3}{2}$

12) $8x+5y=-15$

A) $\frac{8}{5}$ B) $-\frac{8}{5}$

C) $\frac{5}{8}$ D) $-\frac{5}{8}$

13) $x+4y=-8$

A) 4 B) $\frac{1}{4}$

C) -4 D) $-\frac{1}{4}$

14) $x+2y=0$

A) $-\frac{1}{2}$ B) 2

C) -2 D) $\frac{1}{2}$

15) $5x-4y=16$

A) $-\frac{5}{4}$ B) $-\frac{4}{5}$

C) $\frac{5}{4}$ D) $\frac{4}{5}$

Find the slope of a line parallel to each given line.

1) $y = -5$

A) -2 B) Undefined
C) 0 D) 2

2) $5x - 4y = 0$

A) $\frac{5}{4}$ B) $\frac{4}{5}$
C) $-\frac{5}{4}$ D) $-\frac{4}{5}$

3) $x = 1$

A) Undefined B) $-\frac{1}{2}$
C) 0 D) $\frac{1}{2}$

4) $x + 5y = 0$

A) $\frac{1}{5}$ B) 5
C) $-\frac{1}{5}$ D) -5

5) $x - 4y = -20$

A) -4 B) 4
C) $\frac{1}{4}$ D) $-\frac{1}{4}$

6) $x + 4y = -16$

A) 4 B) $-\frac{1}{4}$
C) $\frac{1}{4}$ D) -4

7) $5x - 3y = -9$

A) $-\frac{3}{5}$ B) $-\frac{5}{3}$
C) $\frac{5}{3}$ D) $\frac{3}{5}$

8) $3x + 2y = 4$

A) $-\frac{3}{2}$ B) $-\frac{2}{3}$
C) $\frac{3}{2}$ D) $\frac{2}{3}$

9) $x+5y=15$

A) 5 B) −5

C) $\frac{1}{5}$ D) $-\frac{1}{5}$

10) $x-y=0$

A) 1 B) −1

C) $\frac{5}{3}$ D) $-\frac{5}{3}$

11) $7x+4y=20$

A) $-\frac{4}{7}$ B) $-\frac{7}{4}$

C) $\frac{4}{7}$ D) $\frac{7}{4}$

12) $x=3$

A) Undefined B) $\frac{3}{5}$

C) $-\frac{3}{5}$ D) 0

13) $x-y=1$

A) $\frac{2}{5}$ B) $-\frac{2}{5}$

C) 1 D) −1

14) $x+2y=6$

A) $\frac{1}{2}$ B) $-\frac{1}{2}$

C) 2 D) −2

15) $x-4y=4$

A) 4 B) $-\frac{1}{4}$

C) $\frac{1}{4}$ D) −4

Find the slope of a line perpendicular to each given line.

1) $y = -3x + 2$

A) $\frac{1}{3}$ B) 3

C) -3 D) $-\frac{1}{3}$

2) $y = -\frac{4}{5}x - 3$

A) $\frac{4}{5}$ B) $\frac{5}{4}$

C) $-\frac{5}{4}$ D) $-\frac{4}{5}$

3) $y = -4x + 4$

A) -4 B) $\frac{1}{4}$

C) $-\frac{1}{4}$ D) 4

4) $y = -\frac{3}{2}x - 4$

A) $\frac{3}{2}$ B) $-\frac{2}{3}$

C) $-\frac{3}{2}$ D) $\frac{2}{3}$

5) $y = -\frac{5}{2}x + 2$

A) $\frac{5}{2}$ B) $\frac{2}{5}$

C) $-\frac{5}{2}$ D) $-\frac{2}{5}$

6) $y = x - 1$

A) $-\frac{2}{3}$ B) 1

C) -1 D) $\frac{2}{3}$

7) $y = \frac{6}{5}x + 1$

A) $-\frac{5}{6}$ B) $-\frac{6}{5}$

C) $\frac{6}{5}$ D) $\frac{5}{6}$

8) $y = 3x + 4$

A) 3 B) $\frac{1}{3}$

C) $-\frac{1}{3}$ D) -3

9) $y=-4x$

A) 4 B) $\frac{1}{4}$

C) $-\frac{1}{4}$ D) -4

10) $y=-\frac{4}{5}x+1$

A) $\frac{5}{4}$ B) $-\frac{5}{4}$

C) $-\frac{4}{5}$ D) $\frac{4}{5}$

11) $y=-\frac{1}{3}x+2$

A) 3 B) -3

C) $-\frac{1}{3}$ D) $\frac{1}{3}$

12) $y=\frac{1}{2}x$

A) 2 B) $\frac{1}{2}$

C) $-\frac{1}{2}$ D) -2

13) $y=-\frac{3}{4}x+5$

A) $\frac{4}{3}$ B) $-\frac{3}{4}$

C) $\frac{3}{4}$ D) $-\frac{4}{3}$

14) $y=x+3$

A) $-\frac{1}{2}$ B) 1

C) $\frac{1}{2}$ D) -1

15) $y=\frac{5}{2}x-2$

A) $\frac{2}{5}$ B) $-\frac{2}{5}$

C) $\frac{5}{2}$ D) $-\frac{5}{2}$

Find the value of x for the below questions.

1)

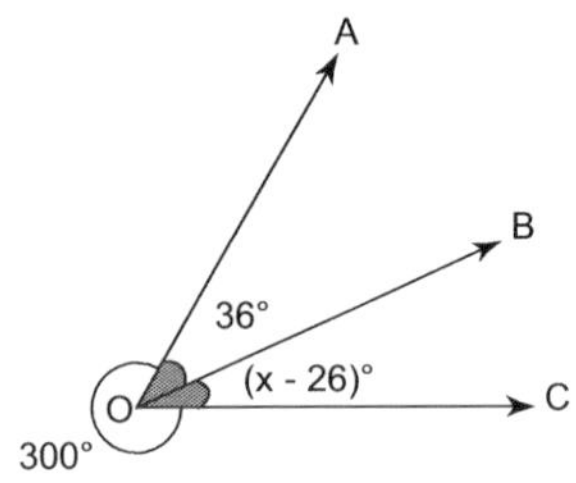

A) 48^0 B) 40^0
C) 50^0 D) 46^0

2)

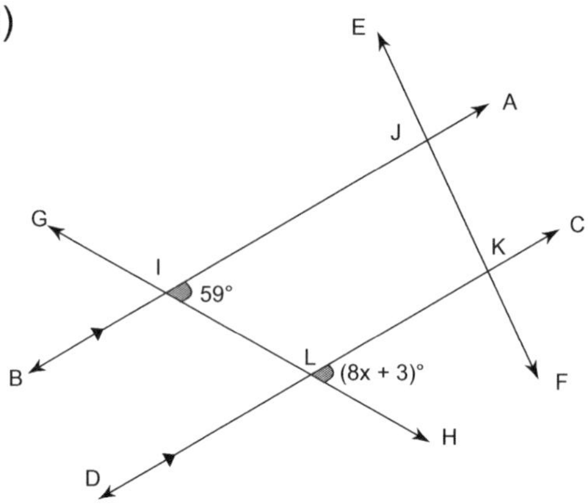

A) 16^0 B) 9^0
C) 12^0 D) 7^0

3)

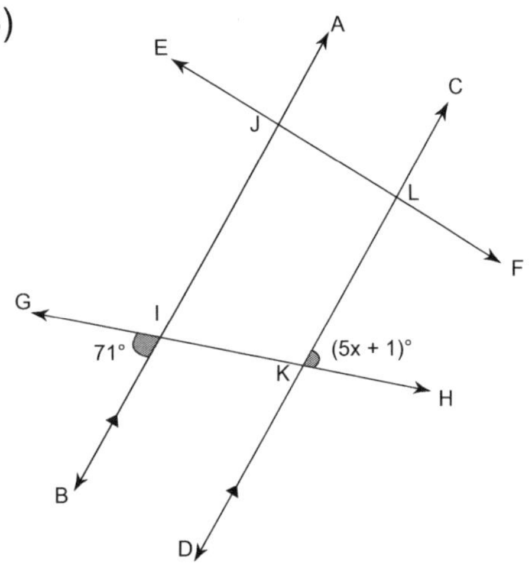

A) -1^0 B) 0^0
C) 14^0 D) 7^0

4)

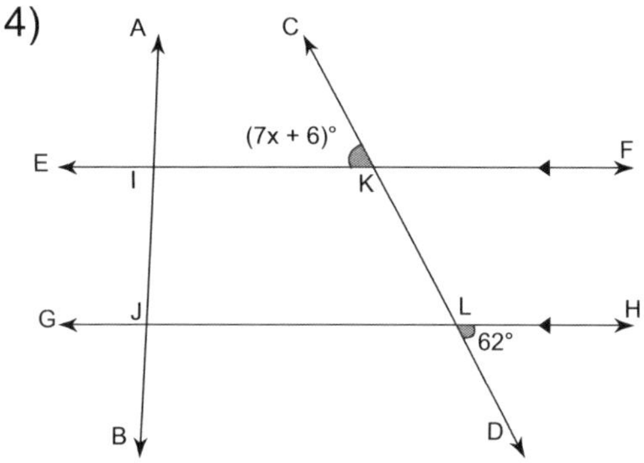

A) 14^0 B) 13^0
C) 8^0 D) 5^0

5)

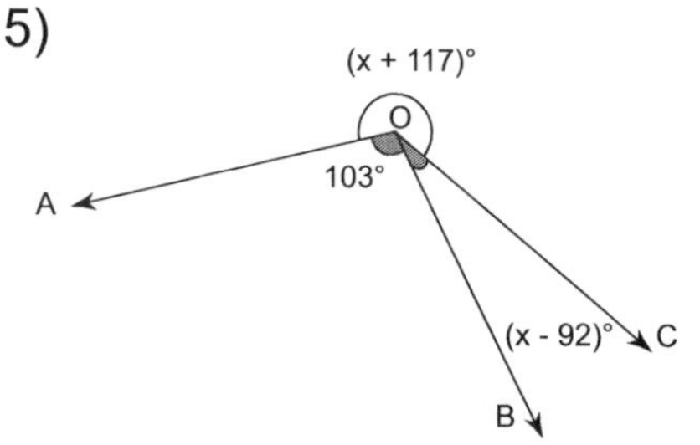

A) 122^0 B) 118^0
C) 115^0 D) 116^0

6)

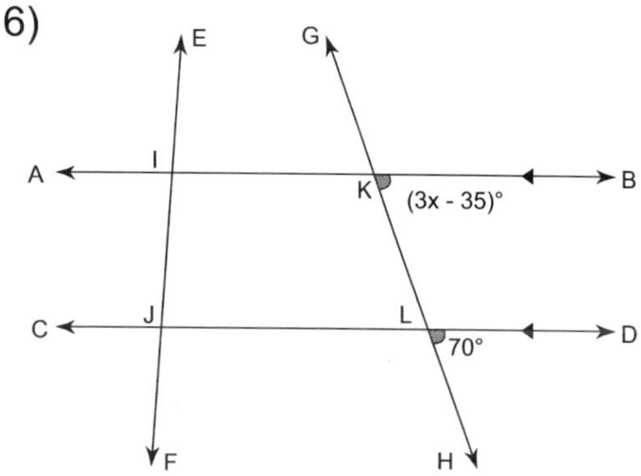

A) 35^0 B) 33^0
C) 31^0 D) 30^0

7)

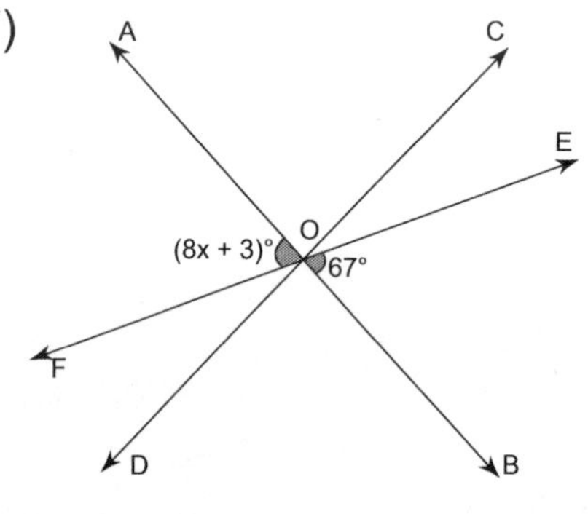

A) 7^0 B) 8^0
C) 15^0 D) 9^0

8)

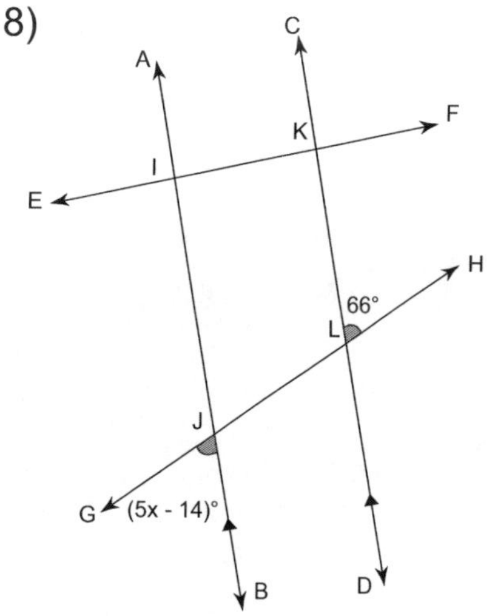

A) 16^0 B) 13^0
C) 21^0 D) 14^0

9)

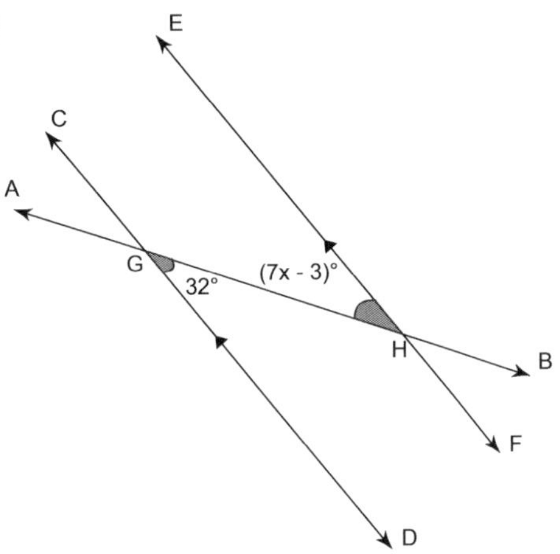

A) 2^0 B) 14^0
C) 5^0 D) 10^0

10)

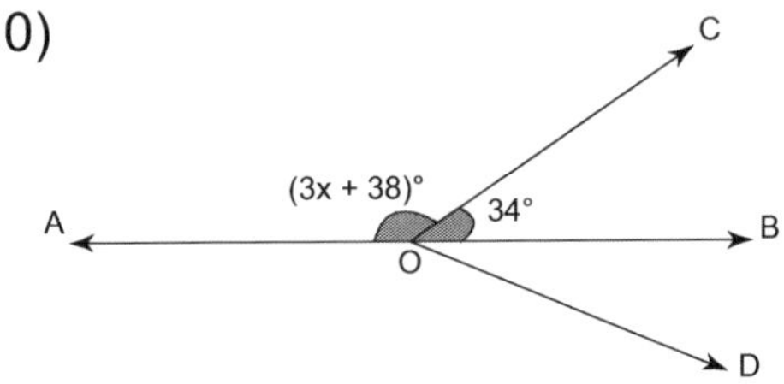

A) 38^0 B) 40^0
C) 36^0 D) 44^0

11)

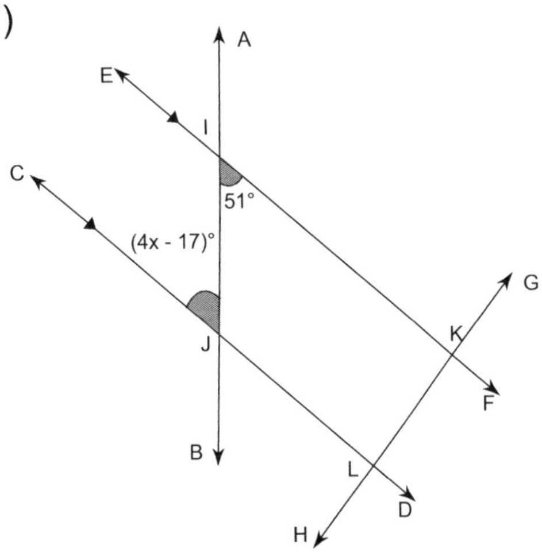

A) 9^0 B) 22^0
C) 25^0 D) 17^0

12)

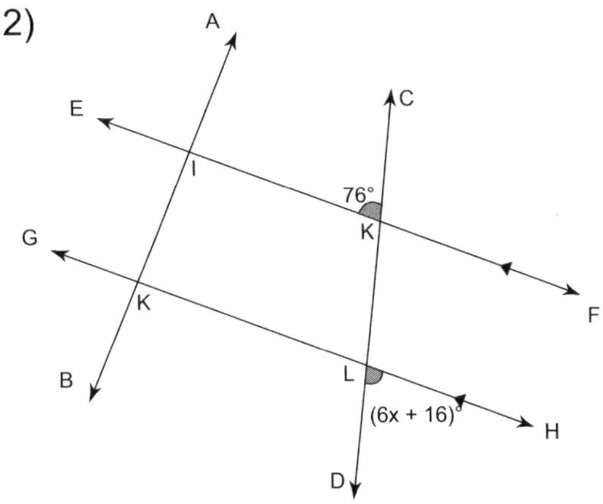

A) 10^0 B) 4^0
C) 11^0 D) -1^0

13)

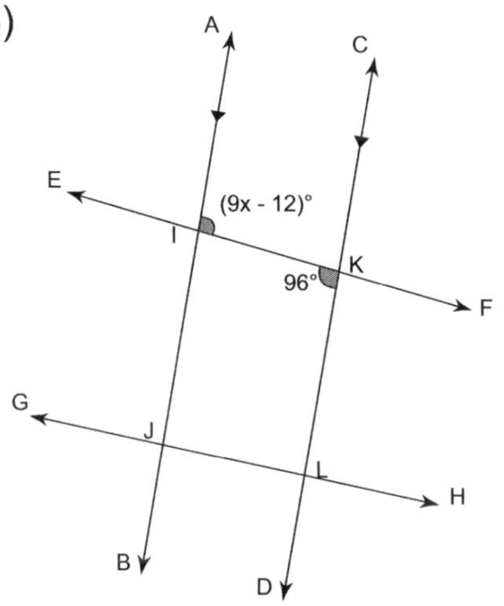

A) 13^0 B) 12^0
C) 11^0 D) 5^0

14)

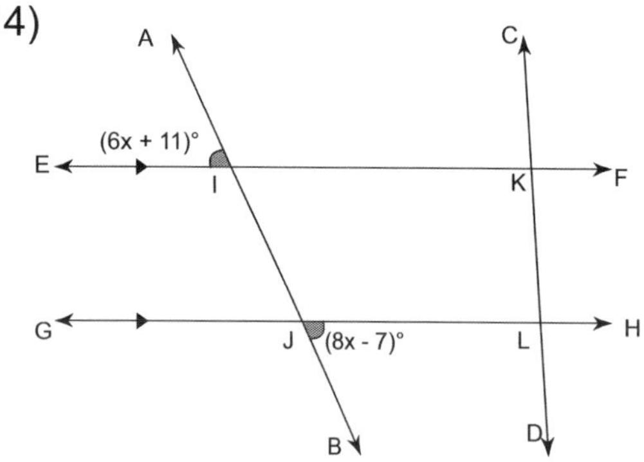

A) 14^0 B) 9^0
C) 18^0 D) 16^0

15)

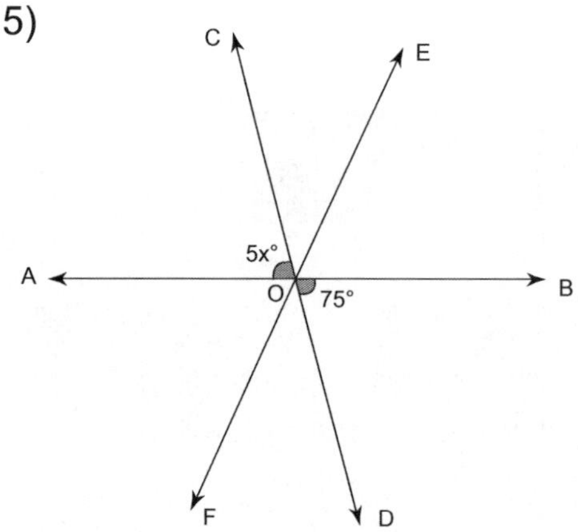

A) 18^0 B) 12^0
C) 6^0 D) 15^0

Find the measure of the ∠x for the below questions:

1)

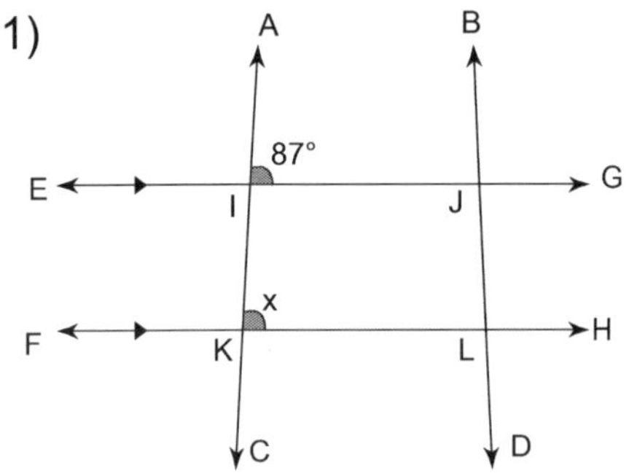

A) 155° B) 87°
C) 93° D) 3°

2)

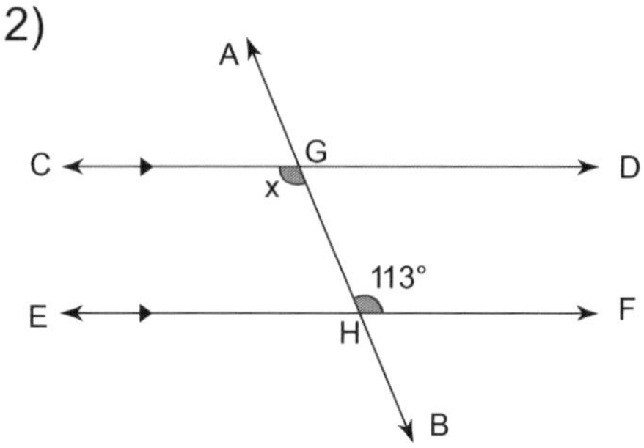

A) 130° B) 67°
C) 50° D) 113°

3)

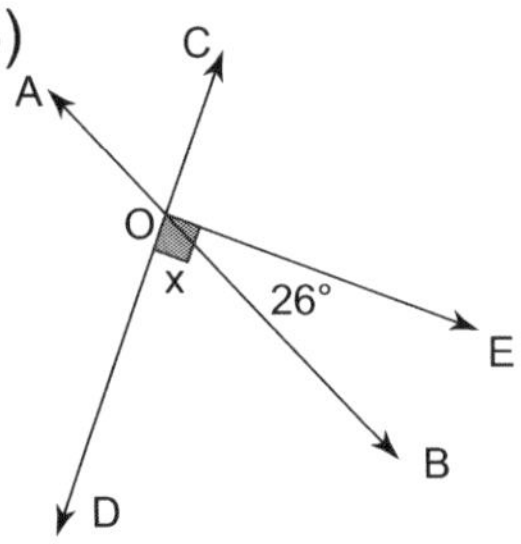

A) 93° B) 154°
C) 26° D) 64°

4)

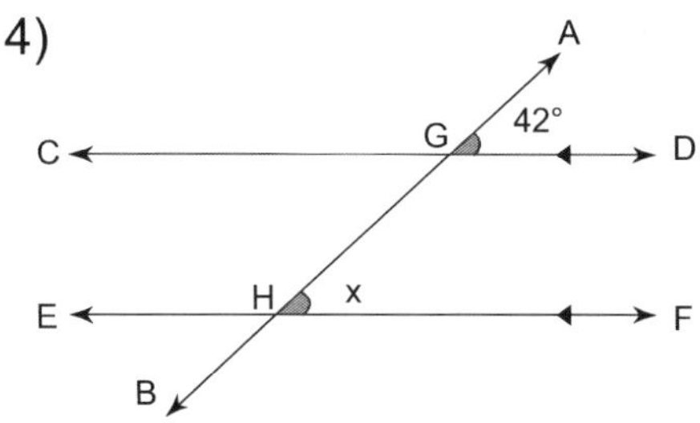

A) 42° B) 132°
C) 138° D) 48°

5)

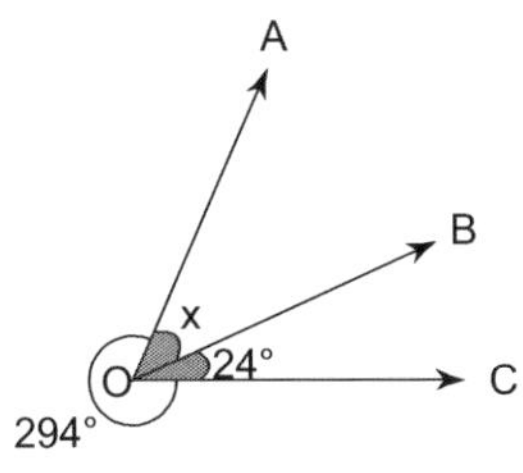

A) 42° B) 132°
C) 48° D) 138°

6)

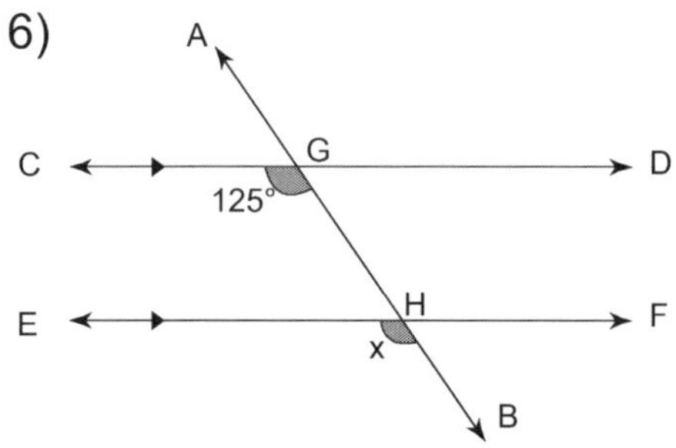

A) 125° B) 35°
C) 145° D) 55°

7)

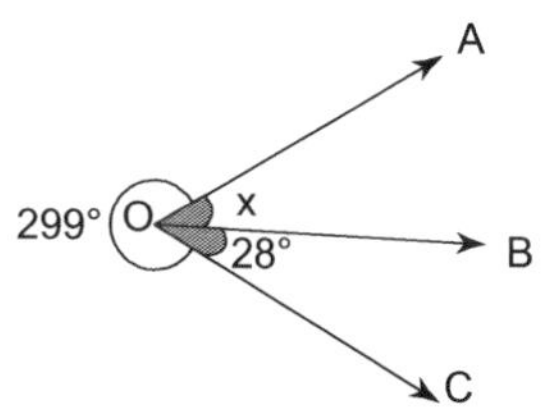

A) 57° B) 147°
C) 33° D) 123°

8)

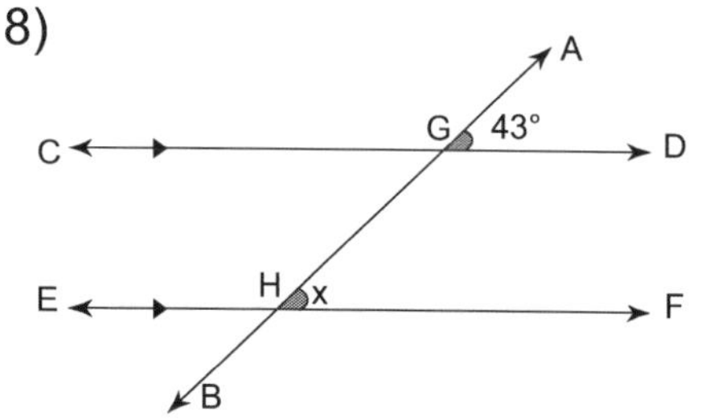

A) 146° B) 43°
C) 137° D) 56°

9)

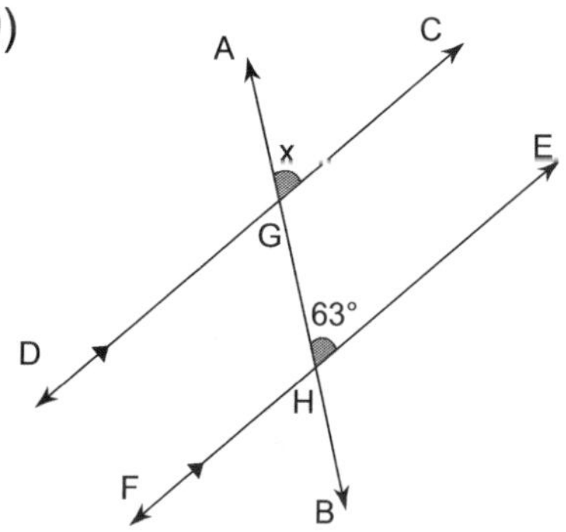

A) 63° B) 87°
C) 117° D) 78°

10)

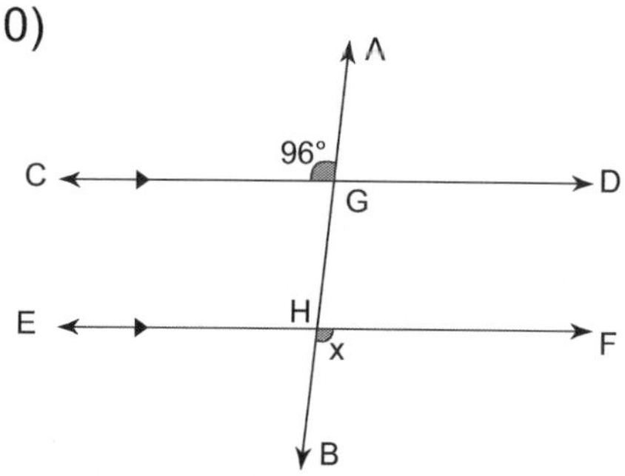

A) 61° B) 84°
C) 96° D) 53°

11)

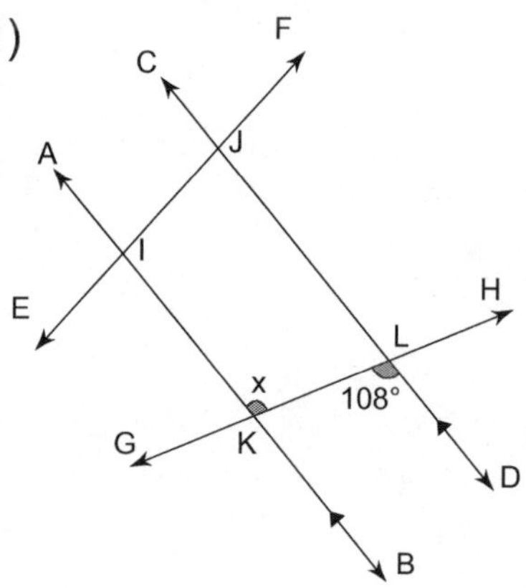

A) 108° B) 18°
C) 162° D) 105°

12)

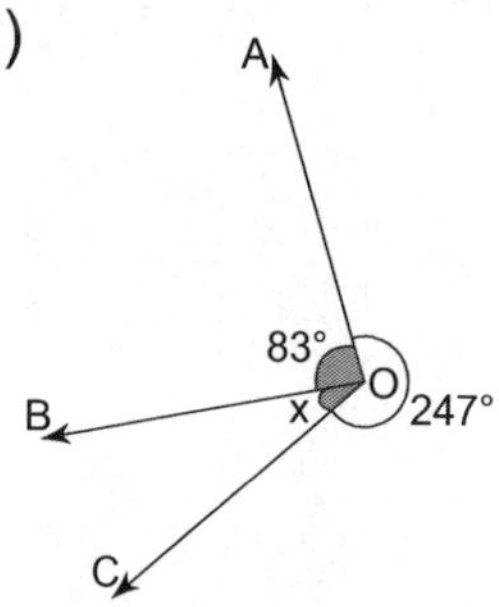

A) 155° B) 30°
C) 150° D) 25°

13)

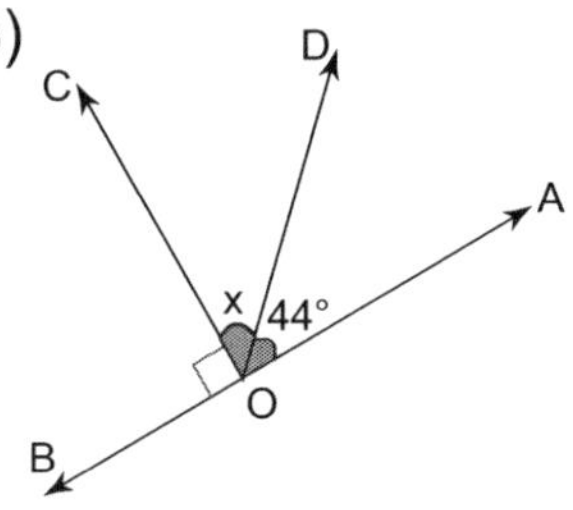

A) 30° B) 134°
C) 46° D) 60°

14)

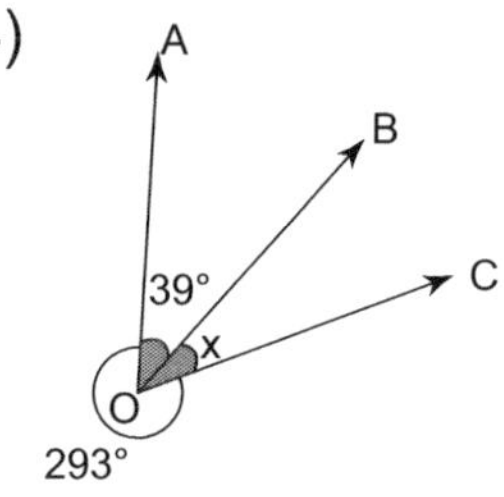

A) 137° B) 28°
C) 133° D) 47°

15)

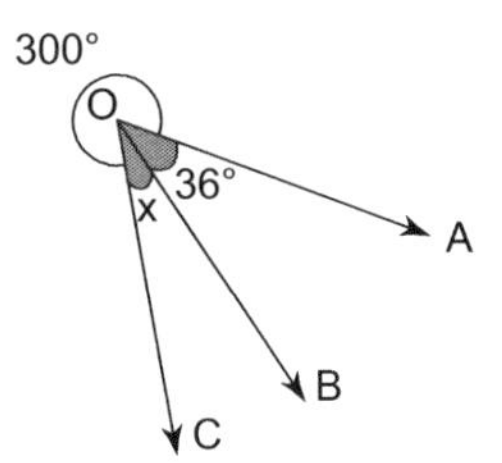

A) 150° B) 156°
C) 24° D) 30°

Describe the transformation and find the transformation rule.

1)

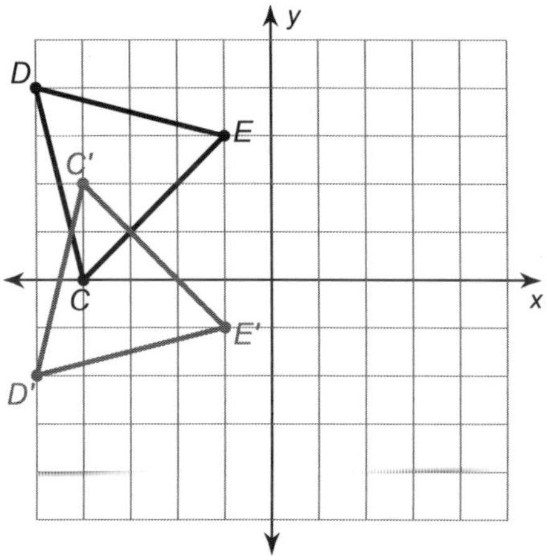

A) Reflection across $y = 1$
B) Reflection across $x = -1$
C) Reflection across $y = -1$
D) Reflection across the x-axis

2)

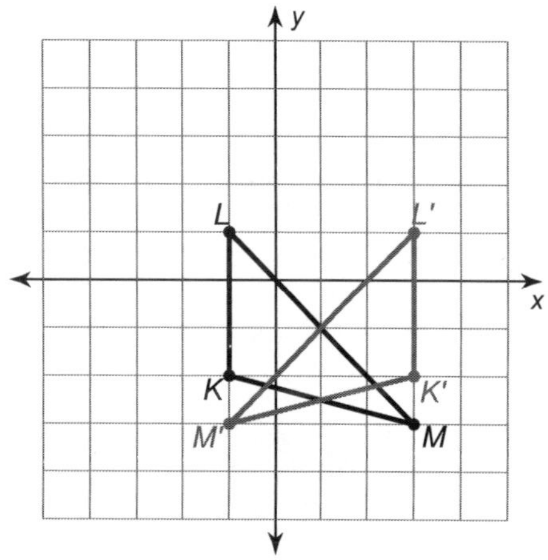

A) Reflection across $x = -2$
B) Reflection across $y = -1$
C) Reflection across $y = 1$
D) Reflection across $x = 1$

3)

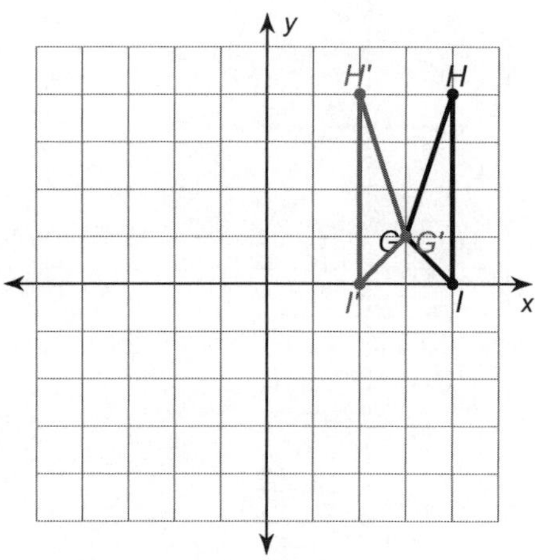

A) Reflection across $x = 1$
B) Reflection across $x = -3$
C) Reflection across $x = -1$
D) Reflection across $x = 3$

4)

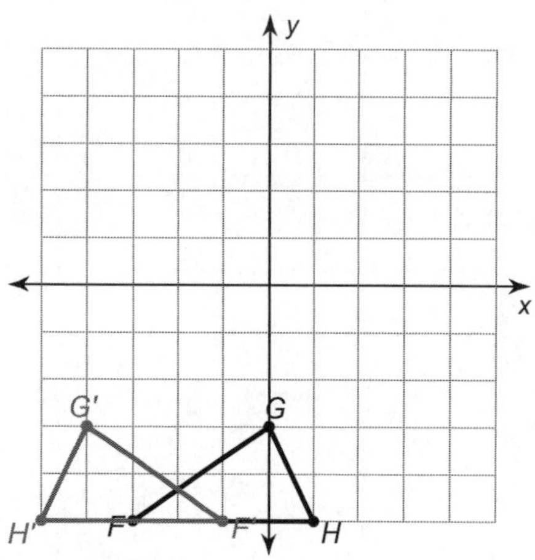

A) Rreflection across the y-axis
B) Reflection across $x = -2$
C) Reflection across $y = -2$
D) Reflection across $y = 1$

Describe the transformation and find the transformation rule.

5)

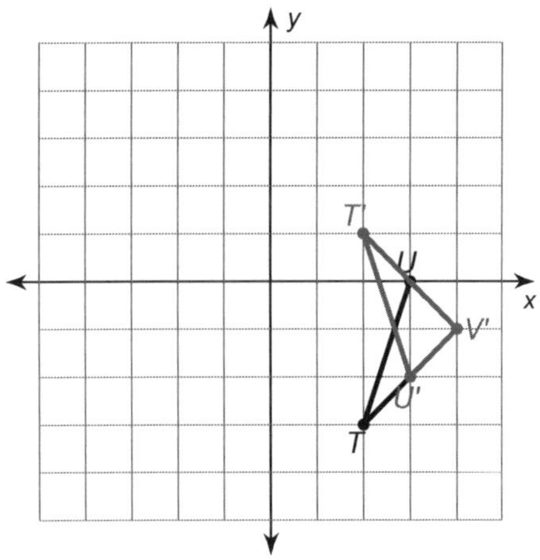

A) Reflection across $y = 1$
B) Reflection across the y-axis
C) Reflection across $x = -1$
D) Reflection across $y = -1$

6)

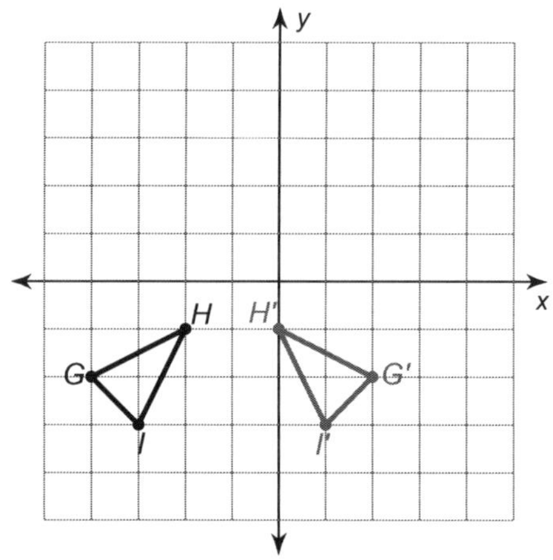

A) Reflection across $x = -1$
B) Reflection across the y-axis
C) Reflection across $x = 1$
D) Reflection across $y = -1$

7)

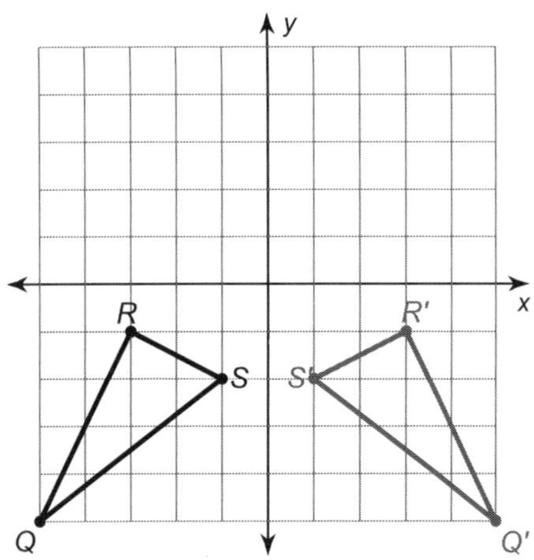

A) Reflection across $x = 1$
B) Reflection across the y-axis
C) Reflection across $y = -1$
D) Reflection across $x = 2$

8)

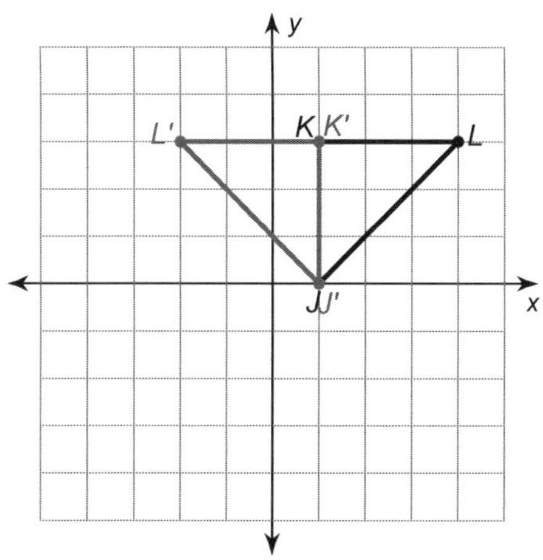

A) Reflection across the x-axis
B) Reflection across $y = 2$
C) Reflection across $x = 1$
D) Reflection across $x = -1$

Describe the transformation and find the transformation rule.

9)

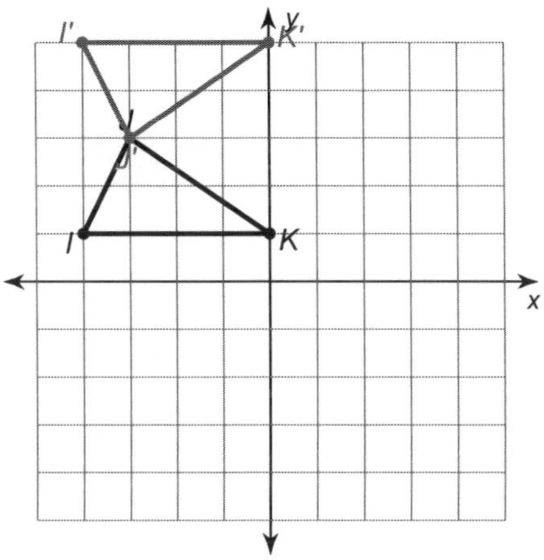

A) Reflection across $x = -3$
B) Reflection across $y = 3$
C) Reflection across $y = -1$
D) Reflection across the y-axis

10)

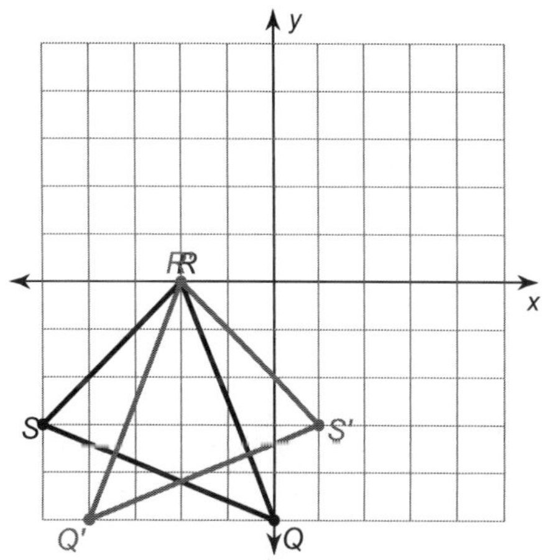

A) Reflection across $y = -2$
B) Reflection across the y-axis
C) Reflection across $x = -2$
D) Reflection across $y = 2$

11)

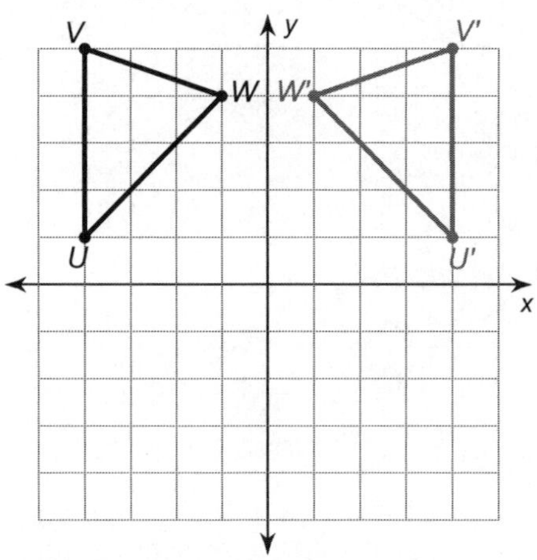

A) Reflection across $x = -1$
B) Reflection across the y-axis
C) Reflection across the x-axis
D) Reflection across $y = -1$

12)

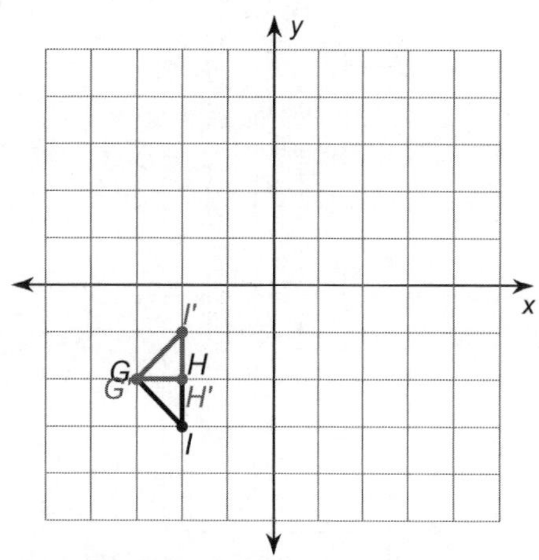

A) Reflection across $y = -2$
B) Reflection across $y = 2$
C) Reflection across $x = -2$
D) Reflection across $x = -1$

Describe the transformation and find the transformation rule.

13)

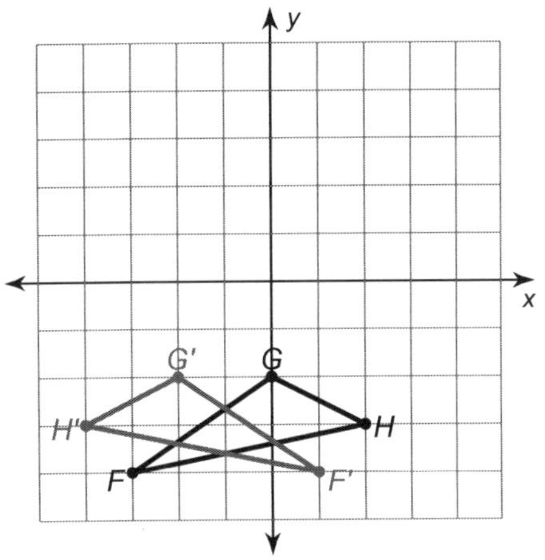

A) Reflection across $x = -1$
B) Reflection across $y = -1$
C) Reflection across x = 1
D) Reflection across the y-axis

14)

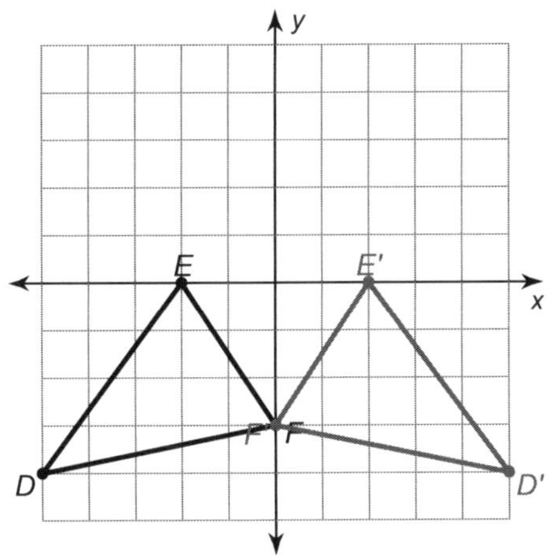

A) Reflection across $y = -4$
B) Reflection across x = 2
C) Reflection across $x = 4$
D) Reflection across the y-axis

15)

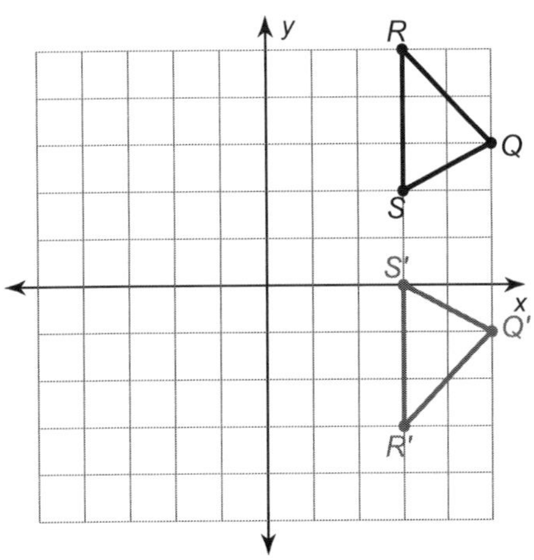

A) Reflection across $y = 1$
B) Reflection across $y = 2$
C) Reflection across y = 8
D) Reflection across $x = 4$

Describe the transformation and find the transformation rule.

1)

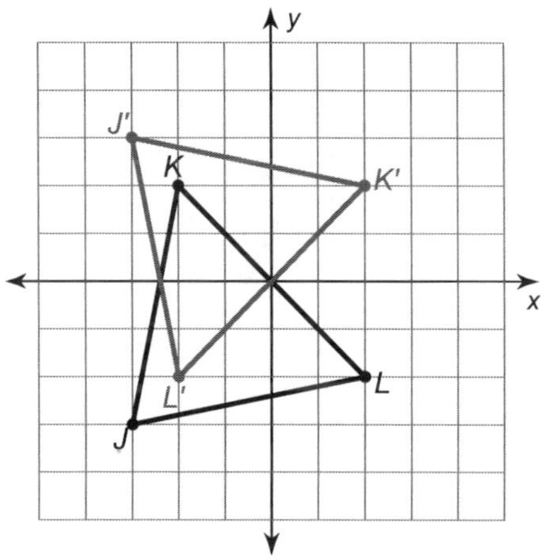

A) Rotation 180° counterclockwise about the origin
B) Translation: 1 unit right and 1 unit up
C) Rotation 90° clockwise about the origin
D) Rotation 270° about the origin

2)

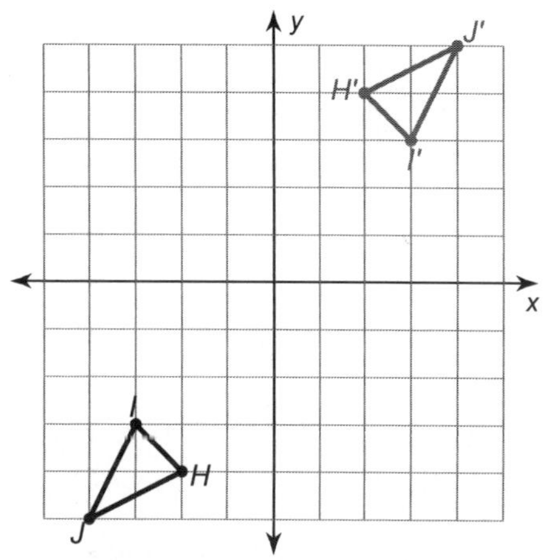

A) Rotation 180° counterclockwise about the origin
B) Rotation 180° about the origin
C) Translation: 8 units right and 5 units up
D) Rotation 90° clockwise about the origin

3)

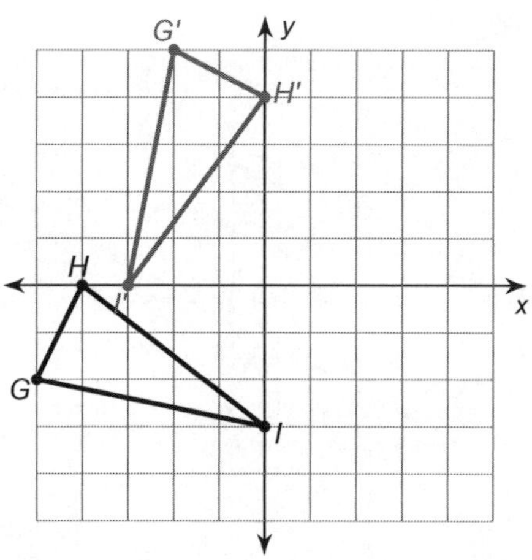

A) Rotation 90° clockwise about the origin
B) Reflection across $y = 4$
C) Rotation 180° counterclockwise about the origin
D) Rotation 90° about the origin

4)

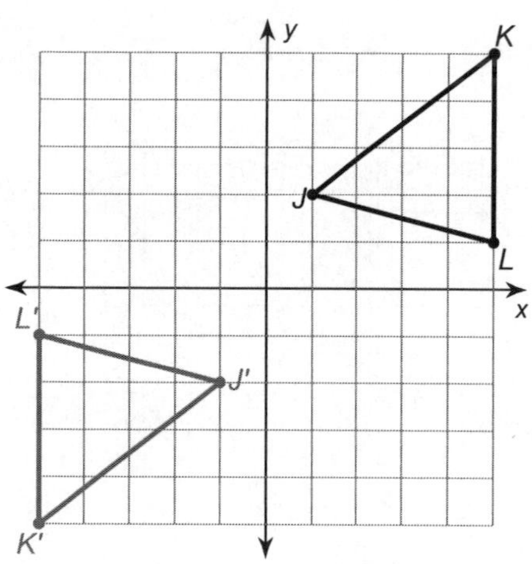

A) Rotation 90° clockwise about the origin
B) Translation: 4 units left
C) Translation: 2 units left
D) Rotation 180° about the origin

Describe the transformation and find the transformation rule.

5)

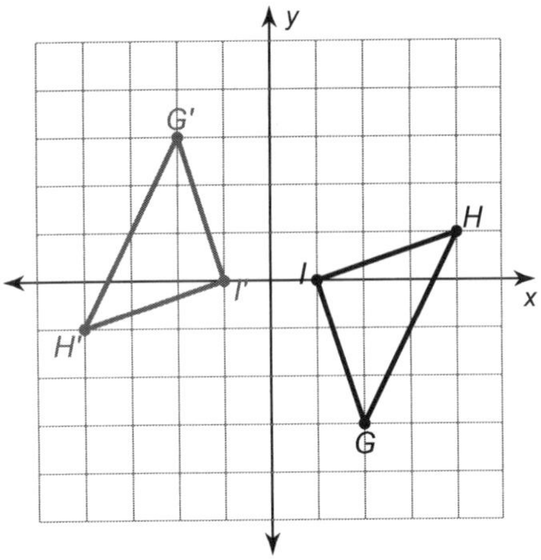

A) Rotation 90° counterclockwise
B) Rotation 180° about the origin
C) Reflection across the y-axis
D) Rotation 90° clockwise about the origin

6)

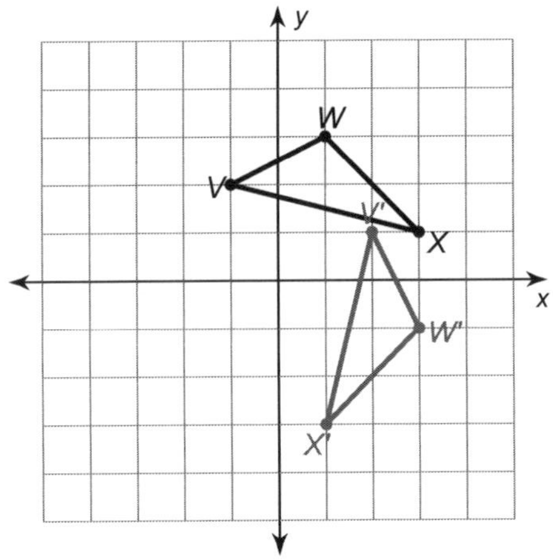

A) Rotation 180° counterclockwise about the origin
B) Translation: 3 units left and 4 units up
C) Rotation 90° clockwise about the origin
D) Rotation 180° about the origin

7)

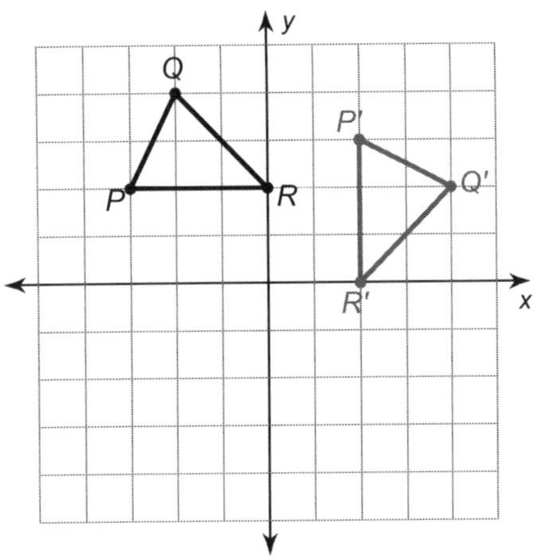

A) Rotation 270° about the origin
B) Rotation 90° clockwise about the origin
C) Rotation 180° counterclockwise about the origin
D) Reflection across $y = 5$

8)

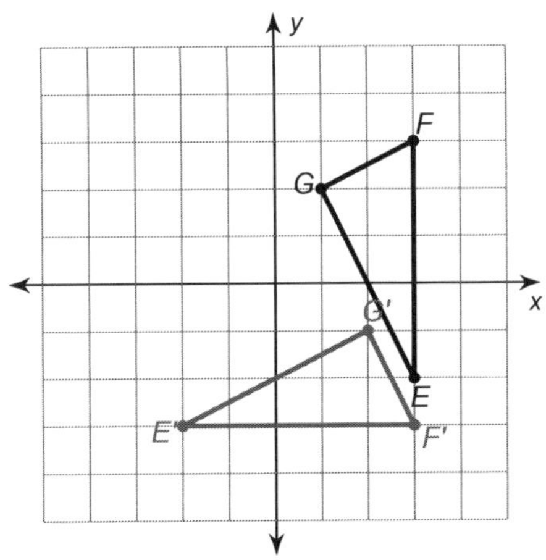

A) Rotation 180° counterclockwise about the origin
B) Translation: 5 units left and 3 units up
C) Rotation 90° clockwise about the origin
D) Rotation 270° about the origin

Describe the transformation and find the transformation rule.

9)

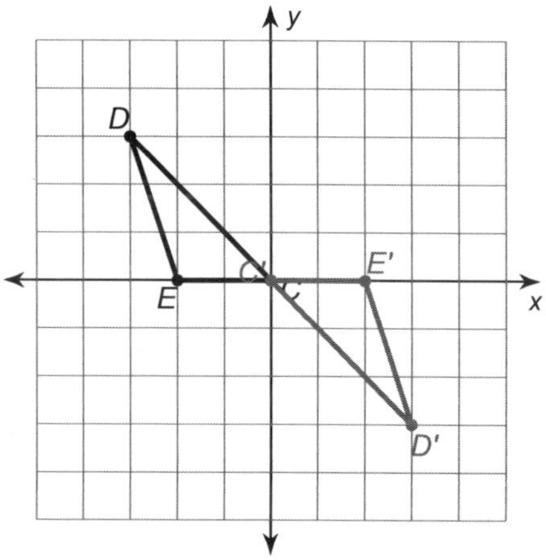

A) Rotation 180° counterclockwise about the origin
B) Rotation 90° clockwise about the origin
C) Translation: 4 units right
D) Rotation 180° about the origin

10)

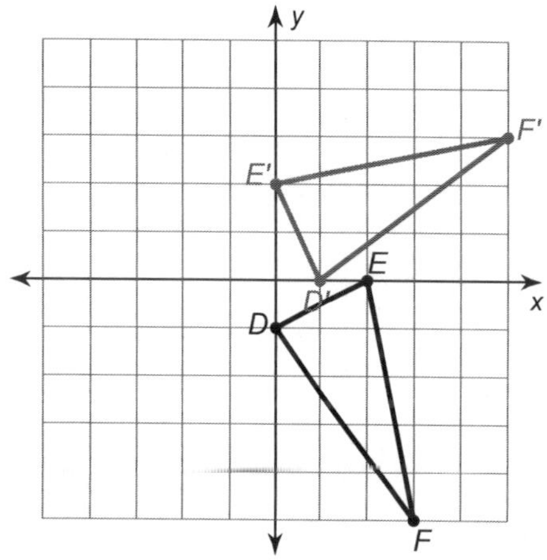

A) Translation: 1 unit right and 3 units up
B) Rotation 180° counterclockwise about the origin
C) Rotation 90° clockwise about the origin
D) Rotation 180° about the origin

11)

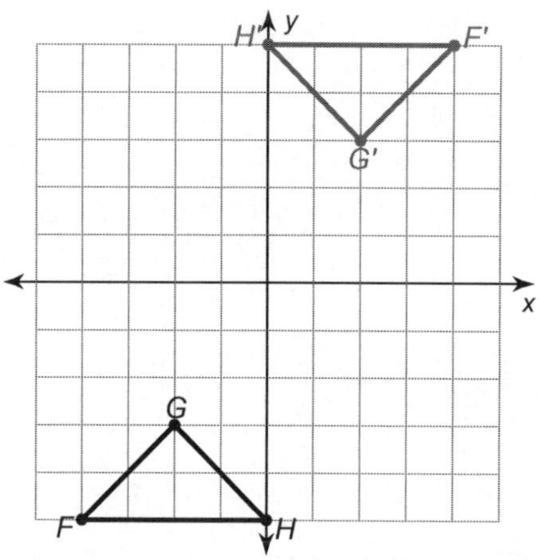

A) Translation: 5 units right and 6 units up
B) Rotation 180° about the origin
C) Rotation 180° counterclockwise about the origin
D) Rotation 90° clockwise about the origin

12)

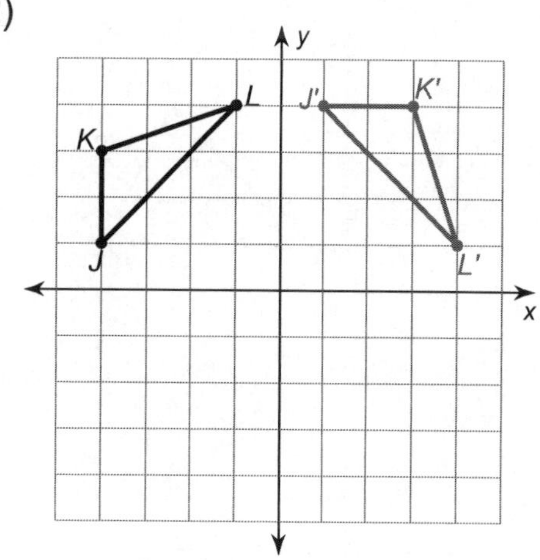

A) Rotation 270° about the origin
B) Rotation 90° clockwise about the origin
C) Translation: 2 units left and 2 units down
D) Rotation 90° counterclockwise about the origin

Describe the transformation and find the transformation rule.

13)

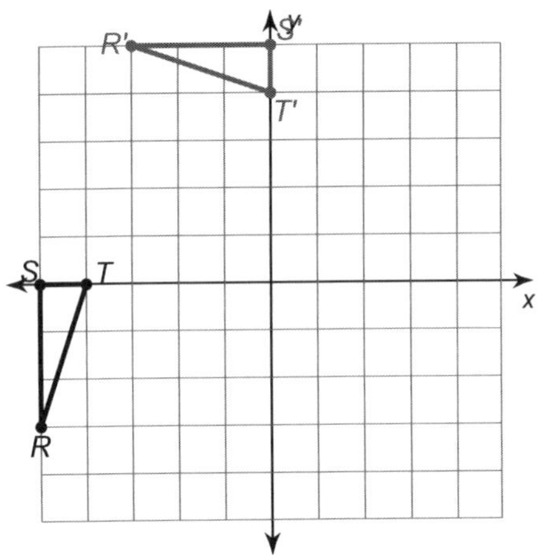

A) Rotation 270° about the origin
B) Reflection across $y = -4$
C) Rotation 180° counterclockwise about the origin
D) Rotation 90° clockwise about the origin

14)

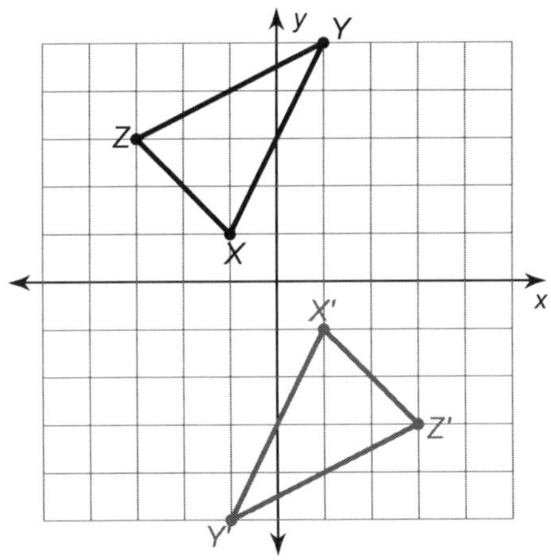

A) Rotation 180° about the origin
B) Reflection across $x = 1$
C) Rotation 90° counterclockwise about the origin
D) Reflection across $x = -1$

15)

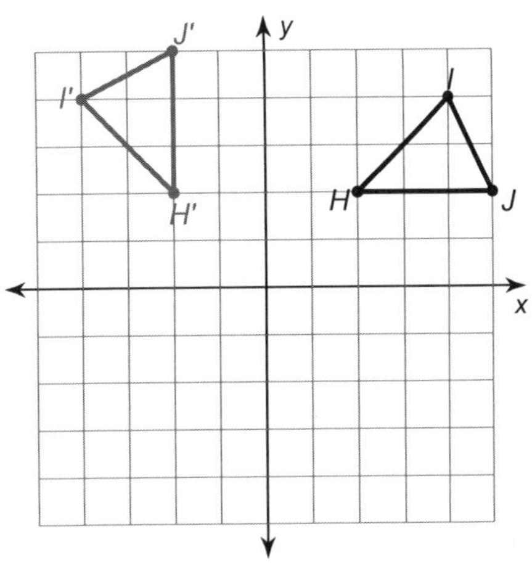

A) Reflection across the y-axis
B) Rotation 90° counterclockwise about the origin
C) Rotation 270° about the origin
D) Rotation 180° clockwise about the origin

Write a rule to describe each transformation.

1)

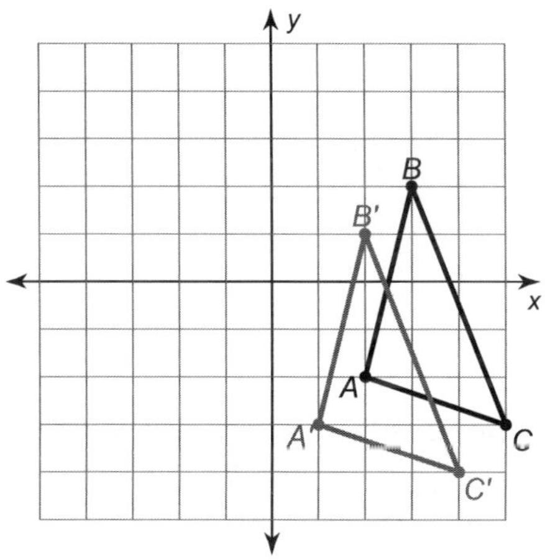

A) Translation: 4 units left and 3 units up
B) Translation: 1 unit left and 1 unit down
C) Translation: 1 unit left and 2 units up
D) Translation: 3 units left and 2 units up

2)

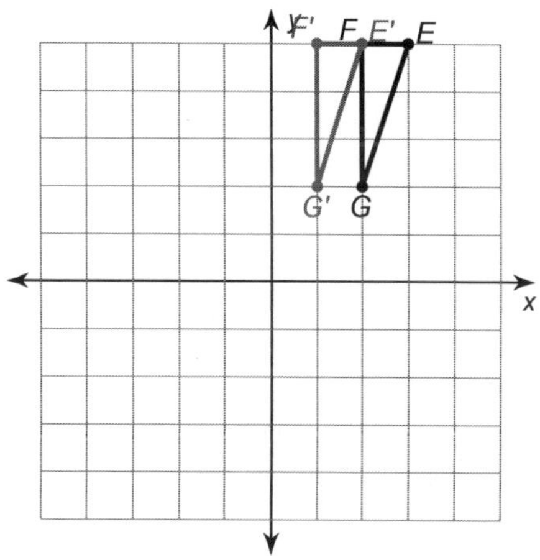

A) Translation: 2 units left and 2 units down
B) Translation: 1 unit right and 1 unit down
C) Translation: 1 unit left
D) Translation: 1 unit right and 2 units down

3)

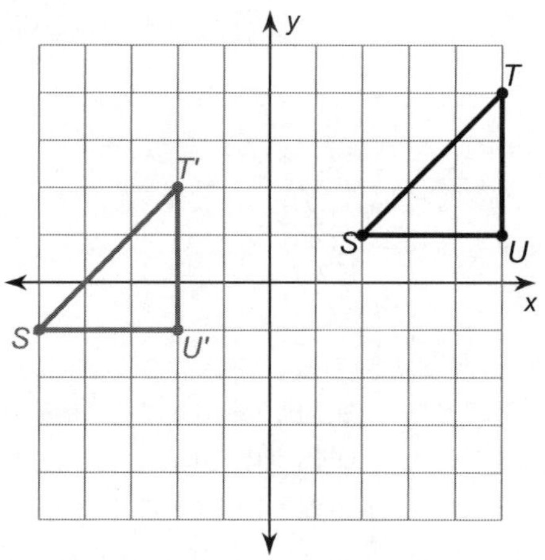

A) Translation: 7 units left
B) Translation: 7 units left and 2 units down
C) Translation: 4 units left and 2 units down
D) Translation: 4 units left and 4 units down

4)

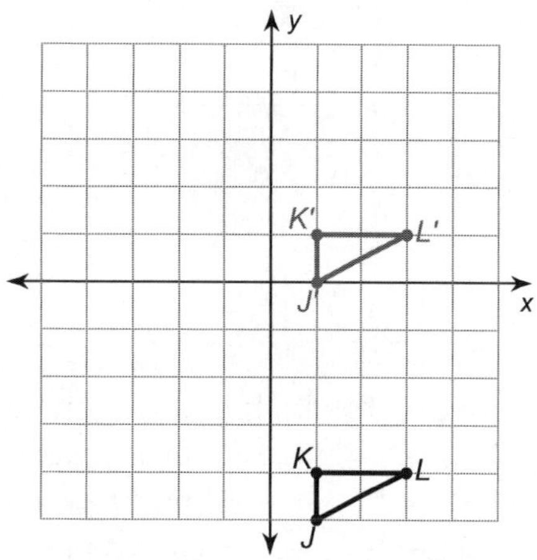

A) Translation: 5 units left and 1 unit down
B) Translation: 5 units up
C) Translation: 2 units left and 3 units up
D) Translation: 1 unit right and 6 units up

5)

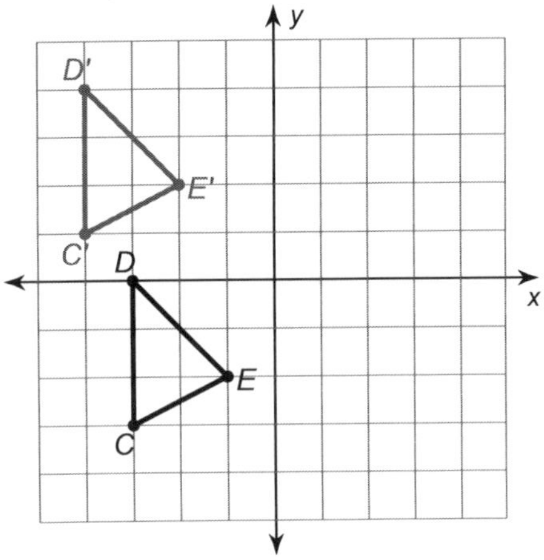

A) Translation: 1 unit left and 4 units up
B) Translation: 4 units right
C) Translation: 1 unit left and 4 units down
D) Translation: 2 units right and 4 units up

6)

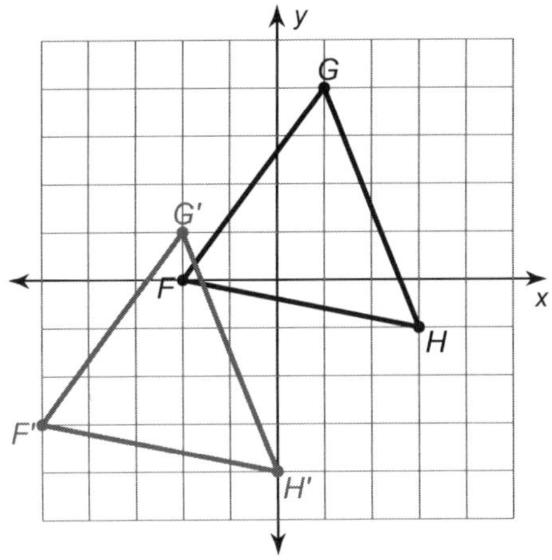

A) Translation: 3 units left and 3 units down
B) Translation: 4 units left and 3 units down
C) Translation: 3 units up
D) Translation: 3 units right

7)

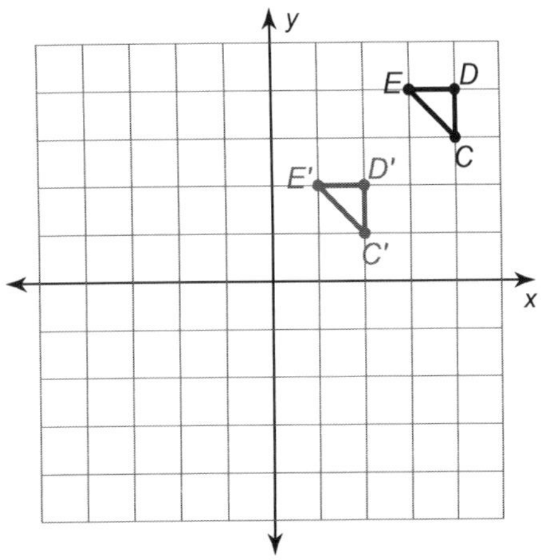

A) Translation: 5 units left and 2 units down
B) Translation: 4 units left and 2 units up
C) Translation: 2 units left and 1 unit down
D) Translation: 2 units left and 2 units down

8)

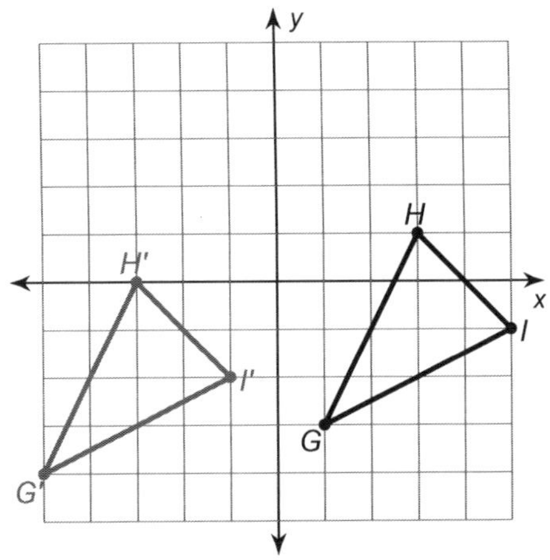

A) Translation: 6 units left and 1 unit up
B) Translation: 1 unit left and 5 units down
C) Translation: 6 units left and 1 unit down
D) Translation: 5 units right

9)

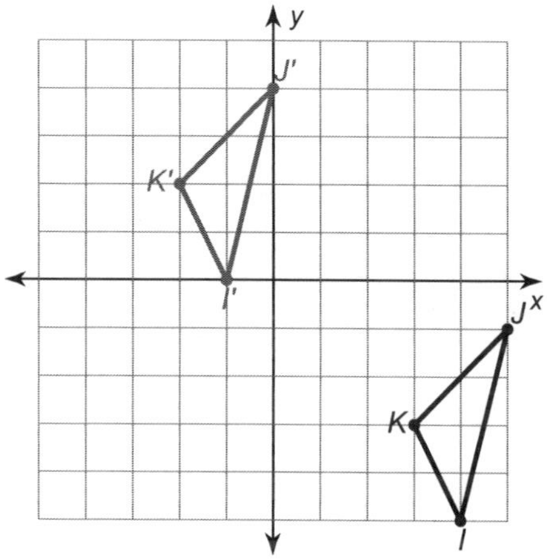

A) Translation: 5 units right and 3 units down
B) Translation: 4 units left and 5 units up
C) Translation: 5 units left and 5 units up
D) Translation: 3 units right and 3 units down

10)

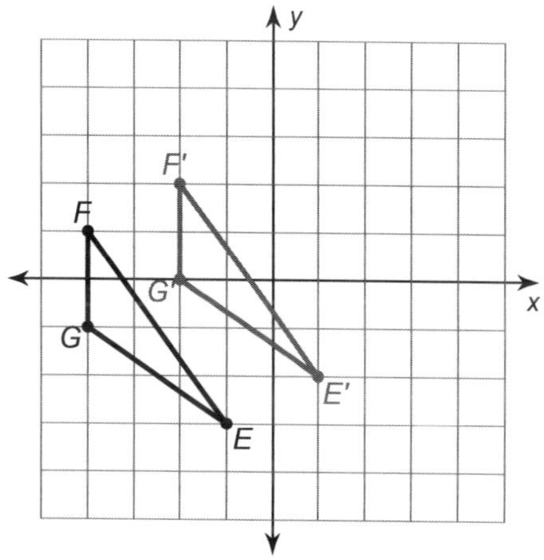

A) Translation: 3 units right
B) Translation: 2 units left
C) Translation: 5 units right and 2 units down
D) Translation: 2 units right and 1 unit up

11)

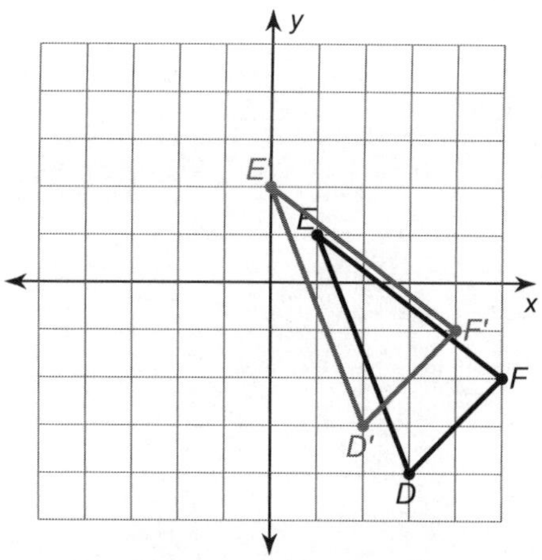

A) Translation: 1 unit right and 2 units down
B) Translation: 1 unit left and 1 unit up
C) Translation: 2 units right
D) Translation: 2 units right and 3 units down

12)

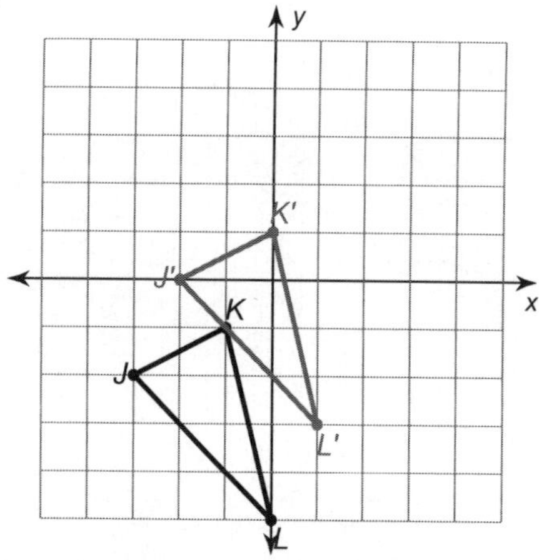

A) Translation: 5 units left
B) Translation: 1 unit right and 2 units up
C) Translation: 3 units right and 2 units down
D) Translation: 3 units left and 2 units up

13)

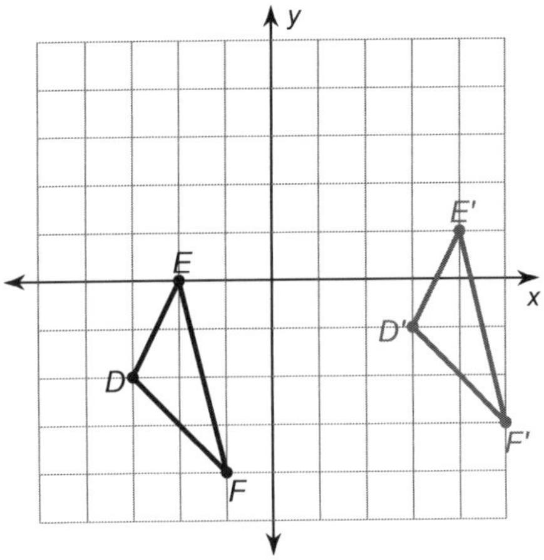

A) Translation: 5 units right and 3 units down
B) Translation: 2 units left and 2 units up
C) Translation: 6 units right and 1 unit up
D) Translation: 7 units left and 1 unit down

14)

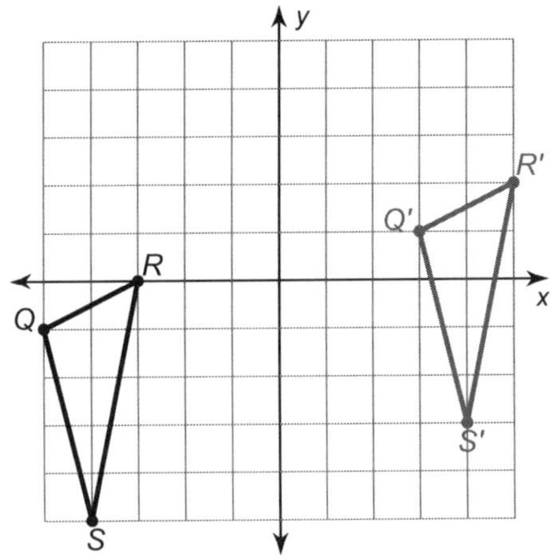

A) Translation: 8 units right and 2 units up
B) Translation: 8 units left
C) Translation: 5 units down
D) Translation: 2 units right and 8 units down

15)

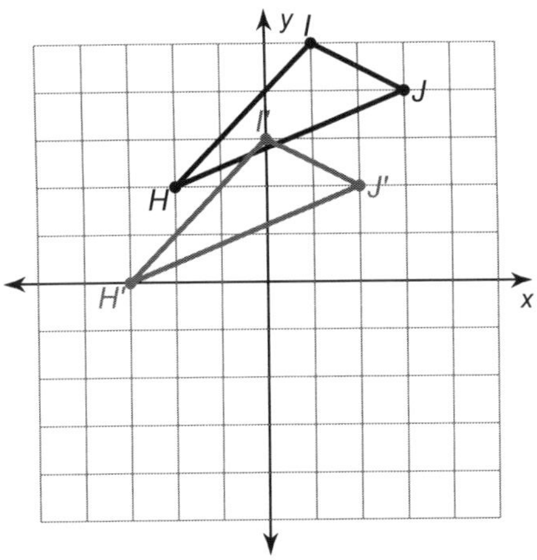

A) Translation: 4 units right and 1 unit up
B) Translation: 3 units up
C) Translation: 1 unit right
D) Translation: 1 unit left and 2 units down

1. Dan was 70 years old when I visited him 9 years ago. How old is he now ?

(A) 79 (B) 61 (C) 75 (D) 88

2. How old is Jason if he was 75 years old thirteen years ago ?

(A) 62 (B) 101 (C) 100 (D) 88

3. Sam spent \$42 on soda packs. If they cost \$6 / pack, how many packs did he buy ?

(A) 7 (B) 12 (C) 11 (D) 10

4. If the weight of a package is multiplied by $\frac{3}{4}$ the result is 61.2 lbs. Find the weight of the package.

(A) 81.6 (B) 61.1 (C) 70.2 (D) 57.4

5. A recipe for pancakes calls for 7 cups of flour. Lucy accidentally put in 10 cups. How many extra cups did she put in ?

(A) 5 (B) 2 (C) 1 (D) 3

6. Mason and his best friend got a cash prize. They divided the money evenly, each getting \$26. How much money was the cash prize ?

(A) \$48 (B) \$52 (C) \$42 (D) \$13

7. Julia and her friends ate 20 muffins. If they ate $\frac{5}{8}$ of what they had originally. How may muffins were there ?

(A) 36 (B) 32 (C) 12.5 (D) 34

8. How old is Claire if she was 30 years old seventeen years ago ?

(A) 47 (B) 13 (C) 64 (D) 54

9. A recipe for a cake calls for $3\frac{2}{7}$ cups of flour. John accidentally put in $4\frac{1}{9}$ cups. How many extra cups did he put in ?

(A) $\frac{207}{259}$ (B) $7\frac{25}{63}$ (C) $4\frac{1}{9}$ (D) $\frac{52}{63}$

10. Susan wants to a gift for her friend's birthday gift. She has \$6 which is $\frac{2}{3}$ of the gift cost. How much did the gift cost ?

(A) \$9 (B) \$5 (C) \$7 (D) \$8

11. Amy wants to buy a pair of shoes for $87. She gives the cashier $90. How much change does she get back ?

(A) $177 (B) $3 (C) $2 (D) $174

12. Jasmine is cooking a casserole. The recipe calls for $3\frac{3}{5}$ cups of rice. She accidentally put in $4\frac{2}{3}$ cups. How many extra cups did she put in ?

(A) $8\frac{4}{15}$ (B) $1\frac{1}{15}$ (C) $4\frac{2}{3}$ (D) $\frac{27}{35}$

13. Cynthia will be 58 years old in ten years. How old is she now ?

(A) 68 (B) 38 (C) 48 (D) 44

14. Dan said, he will be 87 years old in thirteen years. How old is he now ?

(A) 61 (B) 74 (C) 100 (D) 83

15. Kiran ate 12 cookies and realizes he ate $\frac{2}{5}$ of all he had. how many cookies does he had originally ?

(A) 30 (B) 4.8 (C) 26 (D) 24

1. A farmhouse shelters 14 animals. Some are horses and some are geese. Altogether there are 48 legs. How many of each animal are there ?

   (A) 10 geese and 4 horses
   (B) 4 geese and 10 horses
   (C) 12 geese and 2 horses
   (D) 11 geese and 3 horses

2. There are 15 animals in the barn. Some are chickens and some are sheep. There are 50 legs in all. How many of each animal are there ?

   (A) 5 chickens and 10 sheep
   (B) 12 chickens and 3 sheep
   (C) 13 chickens and 2 sheep
   (D) 11 chickens and 4 sheep

3. A farmhouse shelters 23 animals. Some are pigs and some are chickens. Altogether there are 84 legs. How many of each animal are there ?

   (A) 4 chickens and 19 pigs
   (B) 20 chickens and 2 pigs
   (C) 21 chickens and 2 pigs
   (D) 20 chickens and 3 pigs

4. There are 12 animals in the barn. Some are ducks and some are horses. There are 34 legs in all. How many of each animal are there ?

   (A) 8 ducks and 4 horses
   (B) 9 ducks and 3 horses
   (C) 7 ducks and 5 horses
   (D) 10 ducks and 2 horses

5. A farmhouse shelters 26 animals. Some are sheep and some are ducks. Altogether there are 88 legs. How many of each animal are there ?

(A) 22 ducks and 4 sheep
(B) 23 ducks and 3 sheep
(C) 8 ducks and 18 sheep
(D) 24 ducks and 2 sheep

6. There are 12 animals in the barn. Some are geese and some are pigs. There are 42 legs in all. How many of each animal are there ?

(A) 11 geese and 3 pigs
(B) 10 geese and 2 pigs
(C) 9 geese and 3 pigs
(D) 3 geese and 9 pigs

7. A farmhouse shelters 12 animals. Some are cows and some are chickens. Altogether there are 40 legs. How many of each animal are there ?

(A) 8 chickens and 4 cows
(B) 10 chickens and 2 cows
(C) 9 chickens and 3 cows
(D) 4 chickens and 8 cows

8. There are 16 animals in the barn. Some are ducks and some are goats. There are 58 legs in all. How many of each animal are there ?

(A) 13 ducks and 3 goats
(B) 16 ducks and 2 goats
(C) 14 ducks and 2 goats
(D) 3 ducks and 13 goats

9. There are 10 animals in the barn. Some are ducks and some are sheep. There are 30 legs in all. How many of each animal are there ?

(A) 7 ducks and 3 sheep
(B) 9 ducks and 2 sheep
(C) 5 ducks and 5 sheep
(D) 8 ducks and 2 sheep

10. A farmhouse shelters 15 animals. Some are cows and some are chickens. Altogether there are 42 legs. How many of each animal are there ?

(A) 13 chickens and 2 cows
(B) 12 chickens and 2 cows
(C) 9 chickens and 6 cows
(D) 14 chickens and 3 cows

11. There are 15 animals. Some are ducks and some are pigs. There are 52 legs in all. How many of each animal are there ?

(A) 12 ducks and 3 pigs
(B) 14 ducks and 3 pigs
(C) 13 ducks and 2 pigs
(D) 4 ducks and 11 pigs

12. A farmhouse shelters 18 animals. Some are buffalo and some are ducks. Altogether there are 62 legs. How many of each animal are there ?

(A) 5 ducks and 13 buffalos
(B) 15 ducks and 3 buffalos
(C) 17 ducks and 3 buffalos
(D) 16 ducks and 2 buffalos

13. There are 24 animals in the barn. Some are geese and some are buffalo. There are 84 legs in all. How many of each animal are there ?

(A) 6 geese and 18 buffalo
(B) 19 geese and 3 buffalo
(C) 8 geese and 16 buffalo
(D) 22 geese and 2 buffalo

14. There are 27 animals in the barn. Some are geese and some are sheep. There are 94 legs in all. How many of each animal are there ?

(A) 22 geese and 5 sheep
(B) 24 geese and 3 sheep
(C) 25 geese and 2 sheep
(D) 7 geese and 20 sheep

15. A farmhouse shelters 13 animals. Some are pigs and some are chickens. Altogether there are 48 legs in all. How many of each animal are there ?

(A) 9 chickens and 4 pigs
(B) 8 chickens and 5 pigs
(C) 2 chickens and 11 pigs
(D) 11 chickens and 2 pigs

Tracy bought 11 games for a total of $293. Game A cost $25 and Game B cost $28.

1.How many number of game A's did she buy ?

(A) 5 (B) 8

(C) 7 (D) 9

2.How many number of game B's did she buy ?

(A) 2 (B) 3

(C) 4 (D) 6

Jack spent $610 on story books. Fiction books costs $50 and non-fiction books costs $60. He bought a total of 11 story books.

3. How many number of fiction books did he buy ?

(A) 5 Fiction books (B) 9 Fiction books

(C) 8 Fiction books (D) 8 Fiction books

4. How many number of non fiction books did he buy ?

(A) 6 non-fiction books (B) 2 non-fiction books

(C) 3 non-fiction books (D) 2 non-fiction books

Lucy bought 10 dresses for a total of $204. Formal dress cost $26 and casual dress cost $12.

5. How many number of formal dresses did she buy ?

(A) 6 formal dresses (B) 7 formal dresses

(C) 4 formal dresses (D) 7 formal dresses

6. How many number of casual dresses did she buy ?

(A) 6 casual dresses (B) 3 casual dresses

(C) 2 casual dresses (D) 2 casual dresses

Julia spent $40 on snacks. Snack A cost $5 and snack B cost $10. She bought a total of 6 snacks.

7. How many number of snack A did she buy ?

(A) 5 snack A (B) 4 snack A
(C) 2 snack A (D) 3 snack A

8. How many number of snack B did she buy ?

(A) 3 snack B (B) 3 snack B
(C) 4 snack B (D) 2 snack B

Andy bought 7 puzzle sets for a total of $176. puzzle A cost $30 and puzzle B cost $13.

9. How many number of puzzle A's did he buy ?

(A) 4 puzzle A (B) 5 puzzle A
(C) 3 puzzle A (D) 2 puzzle A

10. How many number of puzzle B's did he buy ?

(A) 5 puzzle B (B) 2 puzzle B
(C) 4 puzzle B (D) 3 puzzle B

Kate sold 10 art pieces for a total of $115. Art A cost $8 and Art B cost $15.

11. How many number of Art A pieces did she sold ?

(A) 5 Art A (B) 7 Art A
(C) 8 Art A (D) 3 Art A

12. How many number of Art B pieces did she sold ?

(A) 3 Art B (B) 2 Art B
(C) 5 Art B (D) 4 Art B

Bryan spent $610 on books. Sports books cost $70 and autobiography books cost $80. He bought a total of 8 books.

13. How many number of sports books did she buy ?

(A) 3 sports books
(B) 2 sports books
(C) 6 sports books
(D) 7 sports books

14. How many number of autobiography books did she buy ?

(A) 4 autobiography books
(B) 5 autobiography books
(C) 2 autobiography books
(D) 2 autobiography books

Mary bought 8 plants for a total of $20. Lily plant cost $2 and rose plant cost $3.

15. How many number of Lily plants did she buy ?

(A) 4 Lily plant
(B) 5 Lily plant
(C) 6 Lily plant
(D) 4 Lily plant

Oak middle school took 205 students from grade 6 on a field trip. They were transported in 12 vehicles, in mini vans and buses. Each mini van holds 6 students and each bus hold 25 students.

1. Find the number of mini vans used for transportation.

(A) 5 mini vans (B) 8 mini vans

(C) 10 mini vans (D) 9 mini vans

2. Find the number of buses used for transportation.

(A) 5 buses (B) 4 buses

(C) 2 buses (D) 7 buses

Prim rose elementary took grade 5 students to a science museum. Students are transported in 8 vehicles of cars and buses together. Each car holds 5 students and each bus holds 55 students. A total of 190 students visited the science museum.

3. Find the number of cars used for transportation ?

(A) 6 cars (B) 6 cars

(C) 7 cars (D) 5 cars

4. Find the number of cars used for transportation ?

(A) 4 buses (B) 2 buses

(C) 4 buses (D) 3 buses

220 students from STEM Club participated in a competition this weekend. Few students came in cars in groups of 5 students and other students took buses in groups of 50 students. A total of 8 vehicles were used by students.

5. How many cars were used for transportation ?

(A) 4 cars (B) 6 cars

(C) 5 cars (D) 7 cars

6. How many buses were used for transportation ?

(A) 3 buses
(B) 2 buses
(C) 4 buses
(D) 5 buses

A kindergarten class of 111 students went on a field trip to a pumpkin farm in cars and mini vans. A total of 11 vehicles are used for transportation. Each car carried 5 students and each mini van carried 13 students.

7. How many cars were used for field trip ?

(A) 5 cars
(B) 6 cars
(C) 7 cars
(D) 4 cars

8. How many mini vans were used for field trip ?

(A) 3 mini vans
(B) 2 mini vans
(C) 7 mini vans
(D) 4 mini vans

A class used mini buses and buses to go on a field trip. They used 8 vehicles to go on the trip. Each van holds 10 students and each bus holds 50 students. 280 students went on the trip

9. How many number of mini buses did the class use ?

(A) 8 mini buses
(B) 3 mini buses
(C) 6 mini buses
(D) 4 mini buses

10. How many number of buses did the class use ?

(A) 3 buses
(B) 4 buses
(C) 2 buses
(D) 5 buses

Robotics team of 130 students went to compete in nationals at Washington DC. Students were transported in cars and mini vans. Each car has a capacity of 5 students and each mini van has a capacity of 15 students. A total of 10 vehicles were used for transportation.

11. How many cars were used by Robotics team for transportation?

(A) 6 cars (B) 5 cars
(C) 8 cars (D) 2 cars

12. How many mini vans were used by Robotics team for transportation?

(A) 8 mini vans (B) 5 mini vans
(C) 2 mini vans (D) 3 mini vans

A grade 7 class of 112 students are participating at a county music festival representing their school. Students were transported in 16 vehicles. A group of 5 students traveled in mini cars and groups of nine students traveled in mini vans.

13. How many of mini cars did the class use ?

(A) 7 mini cars (B) 14 mini cars
(C) 8 mini cars (D) 16 mini cars

14. How many of mini vans did the class use ?

(A) 9 mini vans (B) 8 mini vans
(C) 4 mini vans (D) 3 mini vans

A 200 students of Math Club participated in a math bowl competition. Students car pooled in groups of 5 students in cars and groups of 45 students in mini buses. A total of 16 vehicles were used for transportation.

15. How many of cars were used ?

(A) 5 cars (B) 13 cars
(C) 17 cars (D) 14 cars

Based on the below graph answer questions from 1 - 4

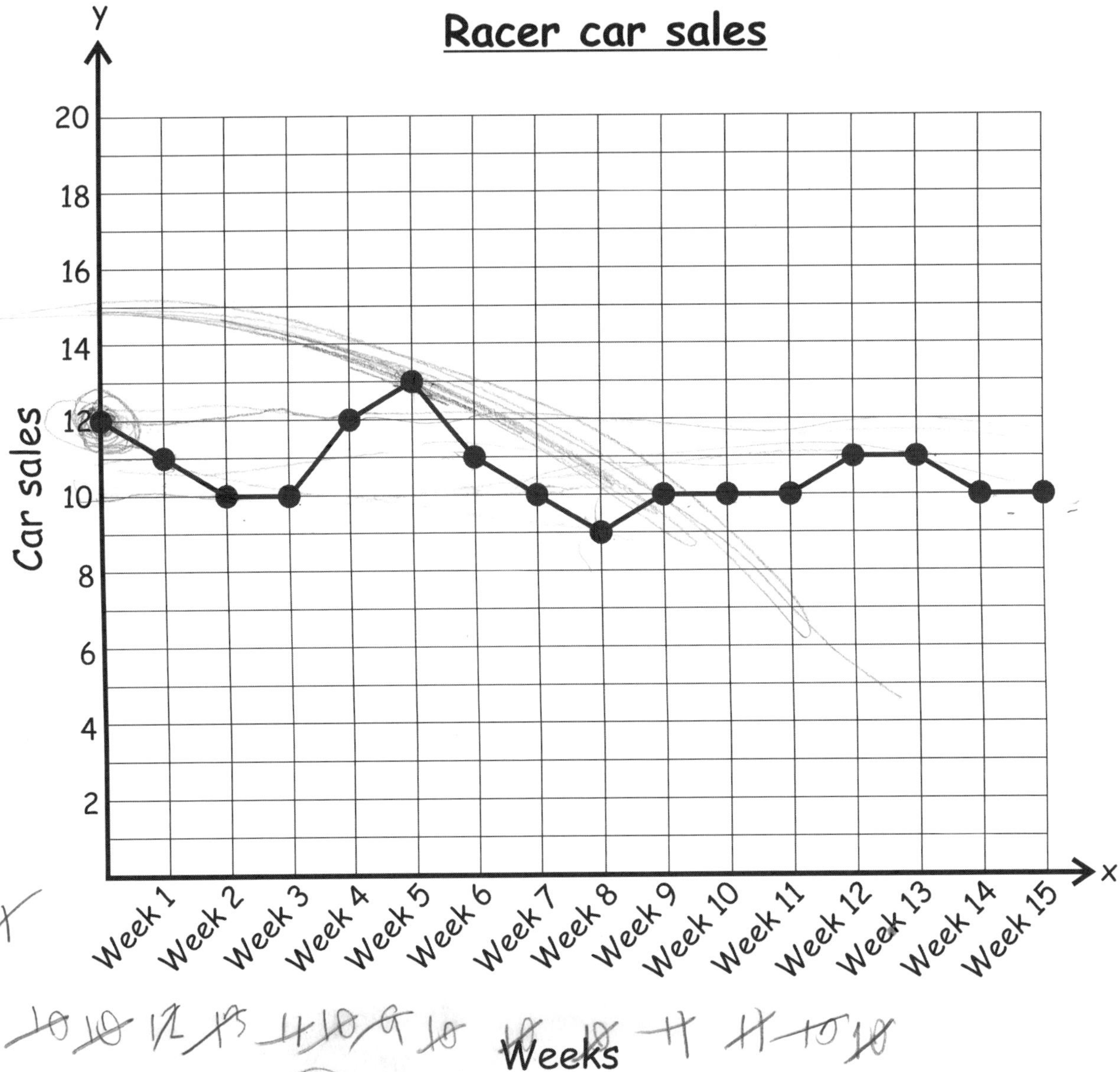

1. Find the median of the above data ?

(A) 16 (B) 12

(C) 13 (D) 10

2. Find the Mean of the above data ?

(A) 10.6 (B) 10.9

(C) 11 (D) 12.5

3. Find the Range of the above data ?

(A) 6 (B) 5

(C) 4 (D) 11

4. What is the Mode of the above data ?

(A) 11 (B) 12

(C) 10 (D) 13

Based on the below graph answer questions from 5 - 8

5. Find the median of the above data ?

(A) 20 (B) 22

(C) 19 (D) 27

6. Find the Mean of the above data ?

(A) 29.1 (B) 22.04

(C) 21.5 (D) 21.4

7. Find the Range of the above data ?

(A) 8 (B) 15

(C) 3 (D) 11

8. What is the Mode of the above data ?

(A) 23 (B) 22

(C) 24 (D) 28

Based on the below graph answer questions from 9 - 12

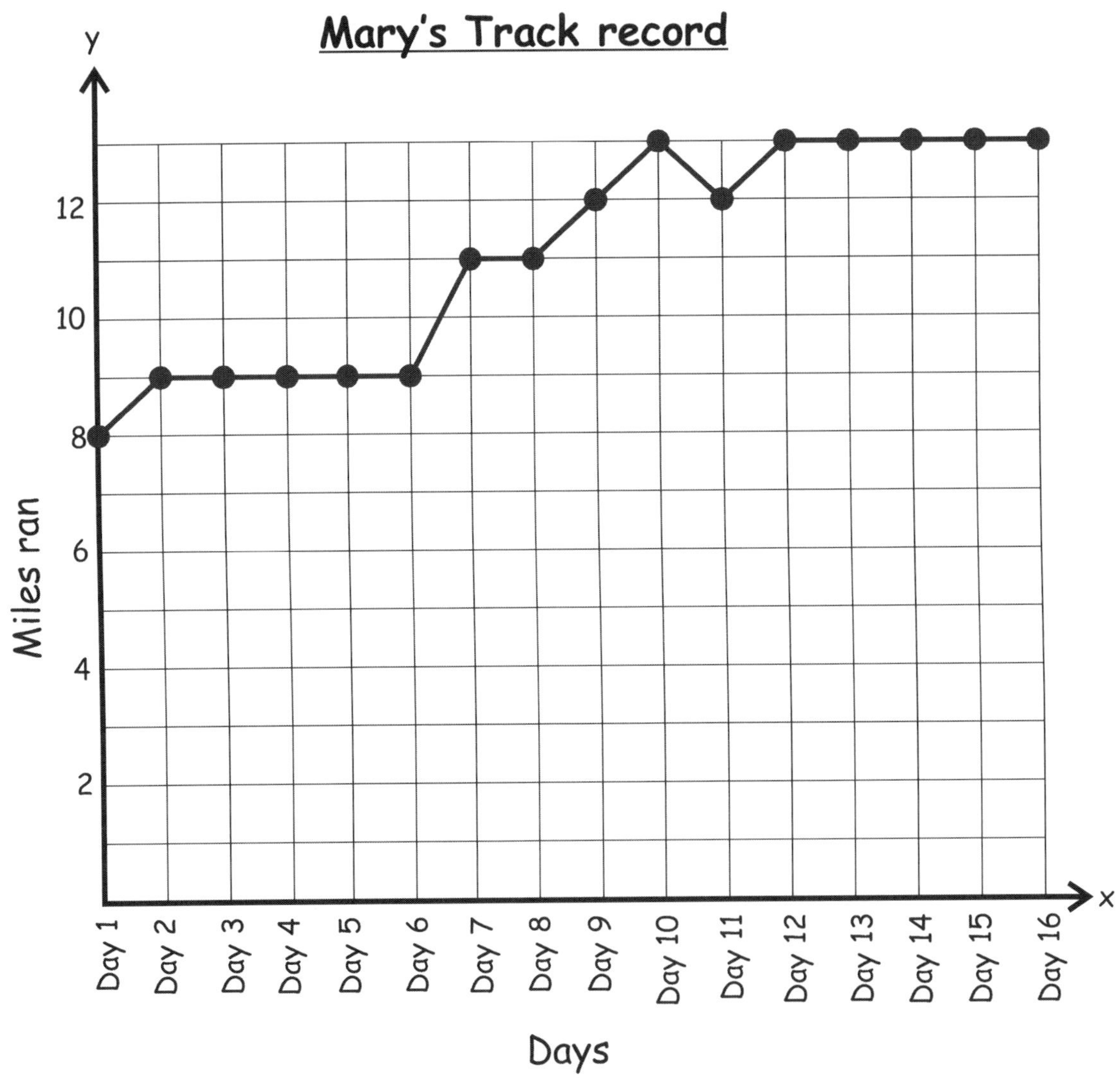

9. Find the median of the above data ?

(A) 11.5 (B) 10

(C) 15 (D) 13

10. Find the Mean of the above data ?

(A) 11.39 (B) 11.09

(C) 11.96 (D) 11.06

11. Find the Range of the above data ?

(A) 6 (B) 5

(C) 9 (D) 14

12. What is the Mode of the above data ?

(A) 13 (B) 12 and 13

(C) 10 (D) 9

Based on the below graph answer questions from 13 - 16

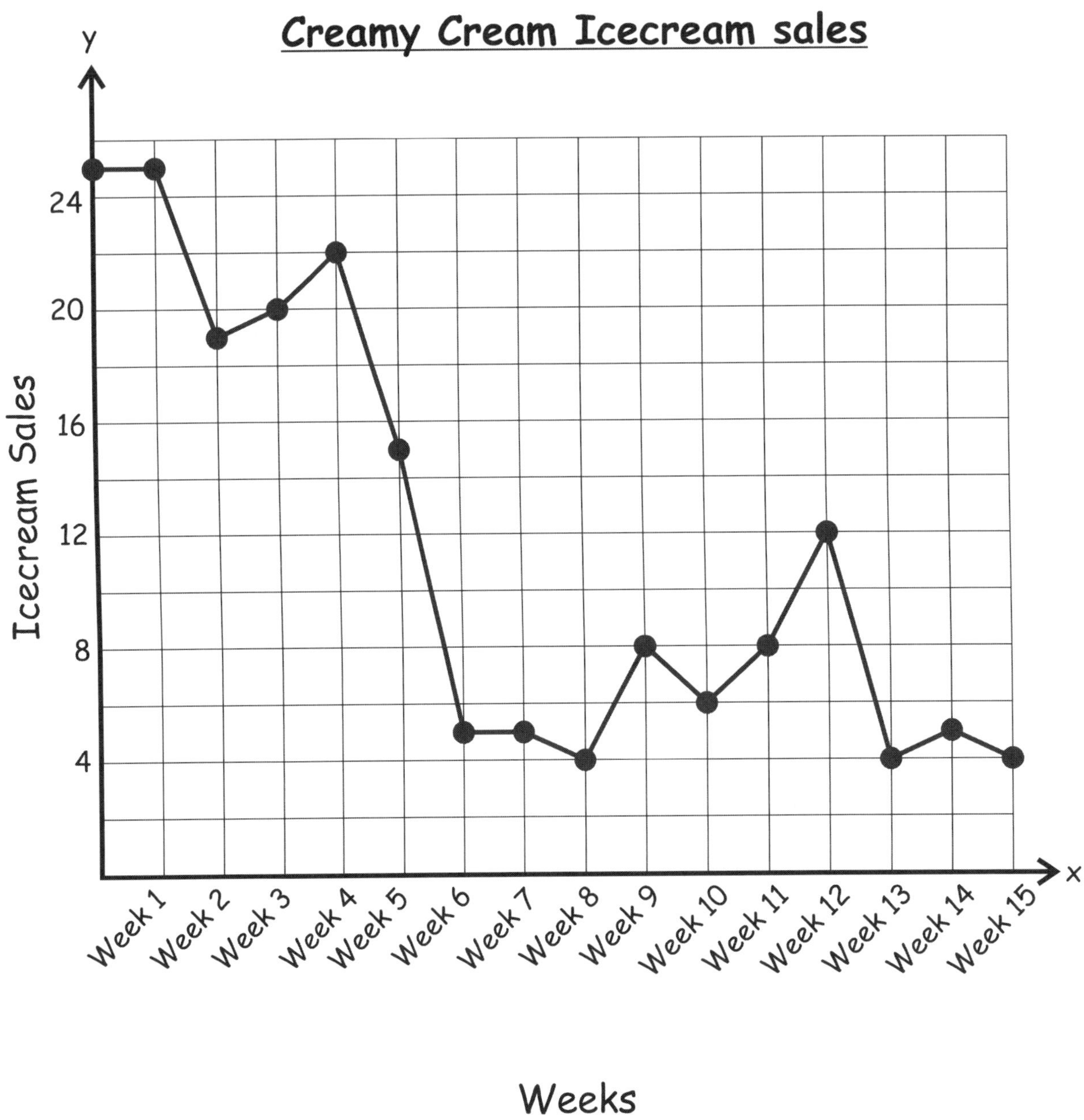

13. Find the median of the above data ?

(A) 15 (B) 8

(C) 11 (D) 10

14. Find the Mean of the above data ?

(A) 11 (B) 12.6

(C) 11.69 (D) 13.9

15. Find the Range of the above data ?

(A) 21 (B) 25

(C) 24 (D) 19

16. What is the Mode of the above data ?

(A) 13 (B) 12

(C) 6 (D) 5

Based on the below graph answer questions from 17 - 20

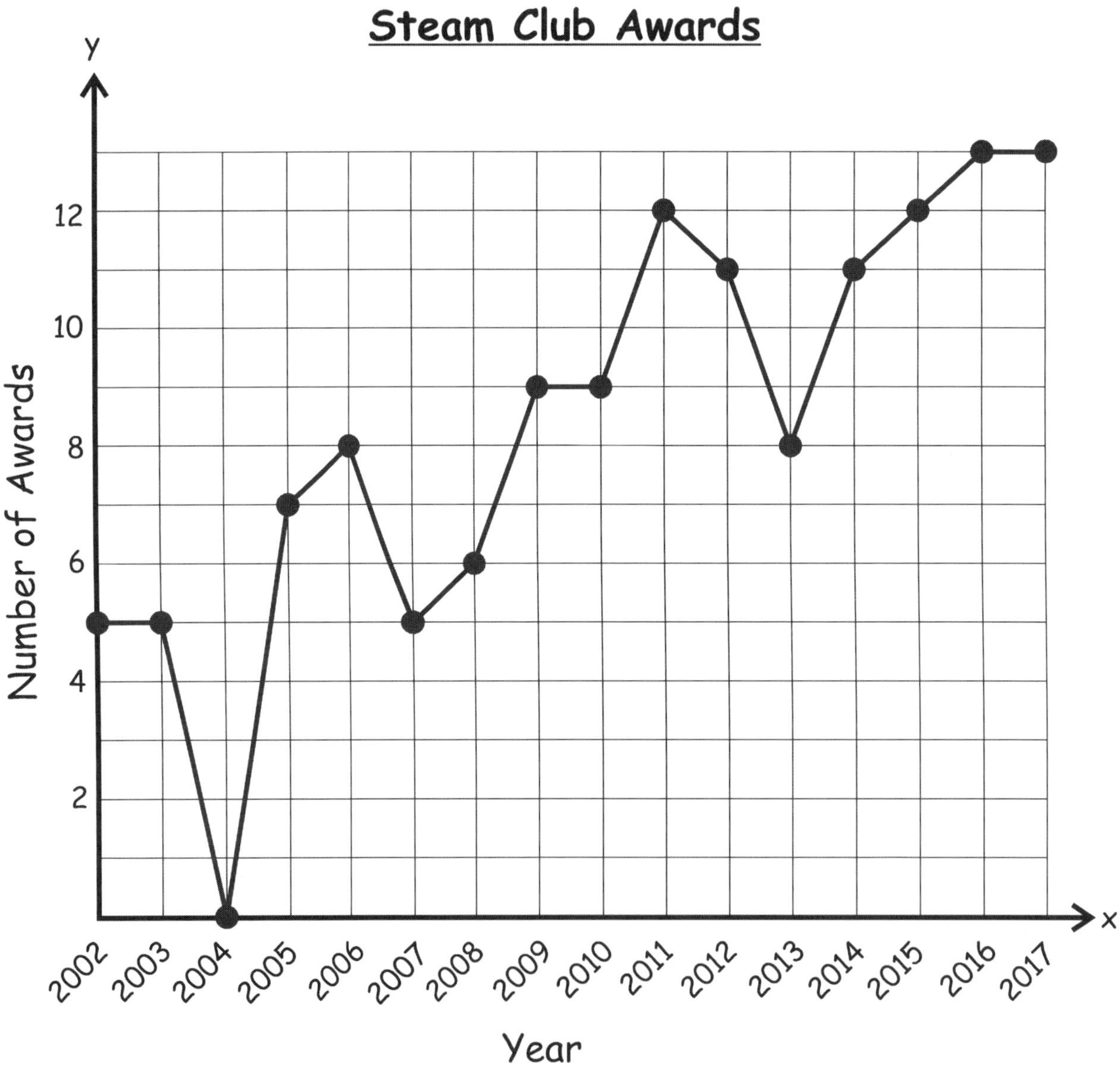

17. Find the median of the above data ?

(A) 8.5 (B) 16

(C) 8.9 (D) 8.1

18. Find the Mean of the above data ?

(A) 7.99 (B) 8.99

(C) 8.52 (D) 8.38

19. Find the Range of the above data ?

(A) 16 (B) 15

(C) 13 (D) 19

20. What is the Mode of the above data ?

(A) 7 (B) 5

(C) 11 (D) 8

1. Find the correlation between the data plotted as below

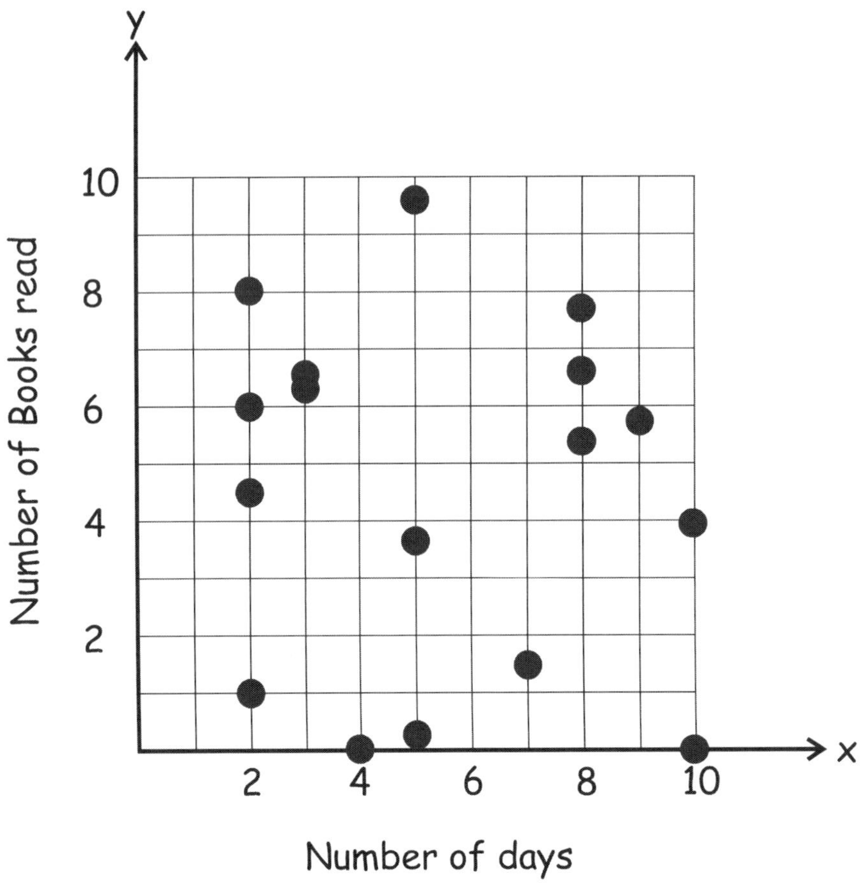

(A) Positive correlation

(B) Negative correlation

(C) No correlation

(D) None

2. Find the correlation between the data plotted as below

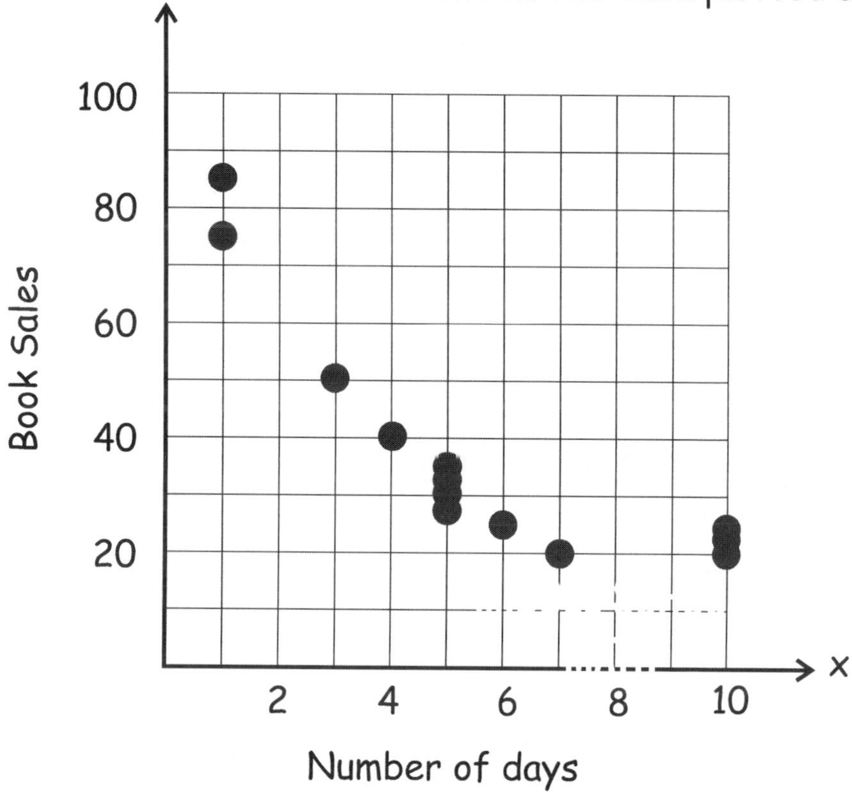

(A) Positive correlation

(B) Negative correlation

(C) No correlation

(D) None

3. Find the correlation between the data plotted as below

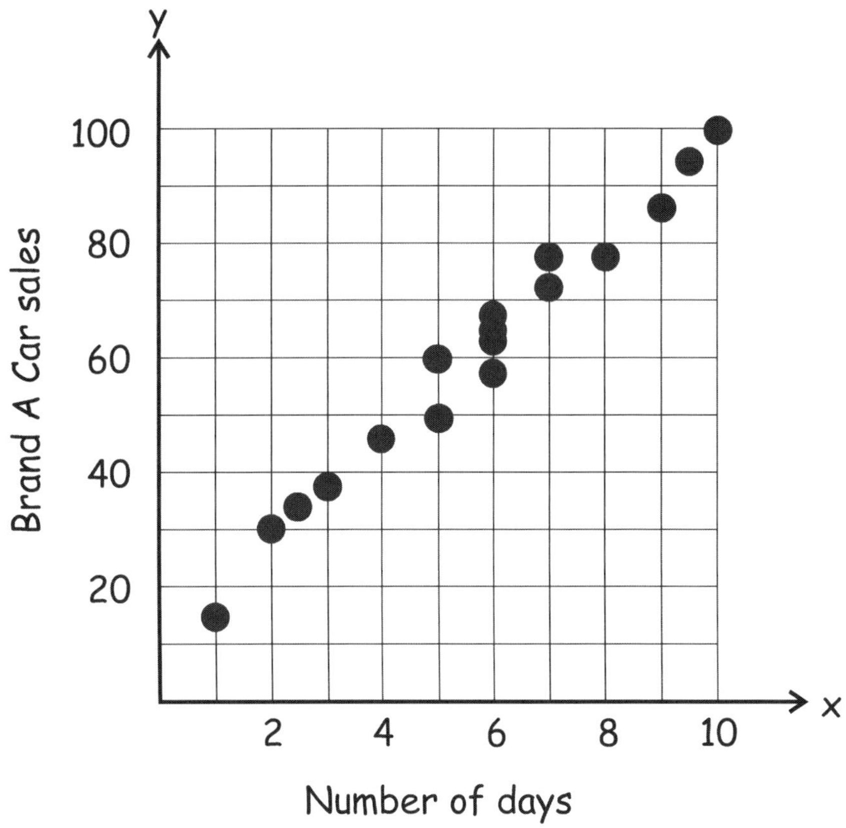

(A) Positive correlation

(B) Negative correlation

(C) No correlation

(D) None

4. Find the correlation between the data plotted as below

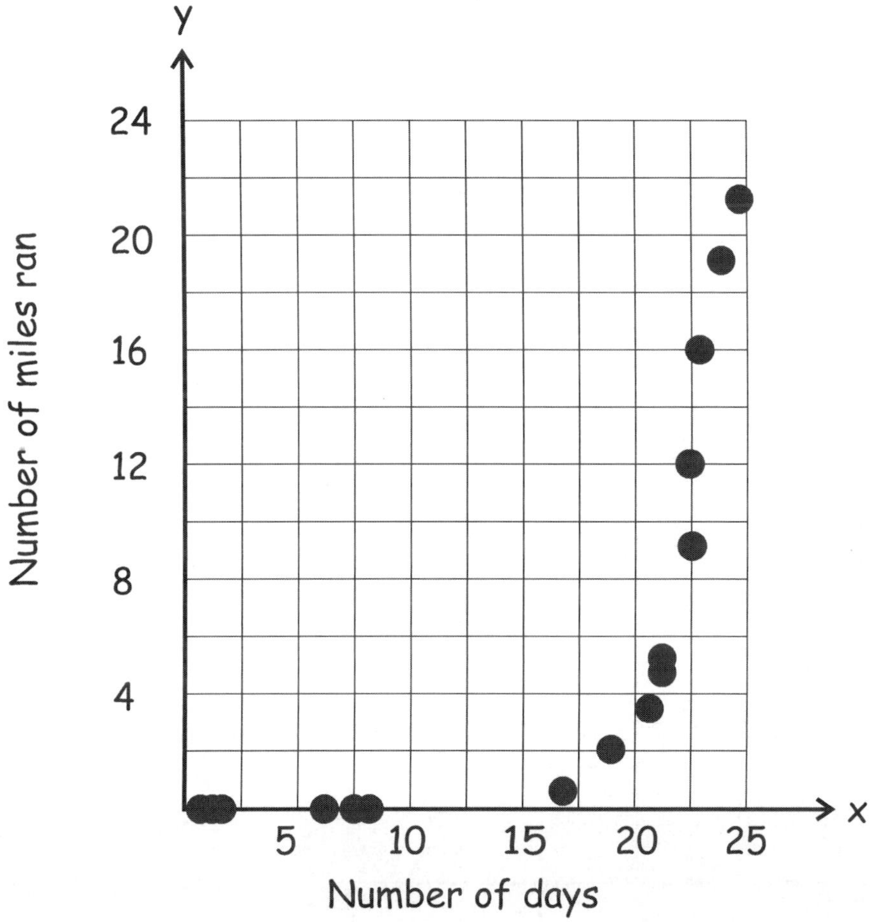

(A) Positive correlation

(B) Negative correlation

(C) No correlation

(D) None

5. Find the correlation between the data plotted as below

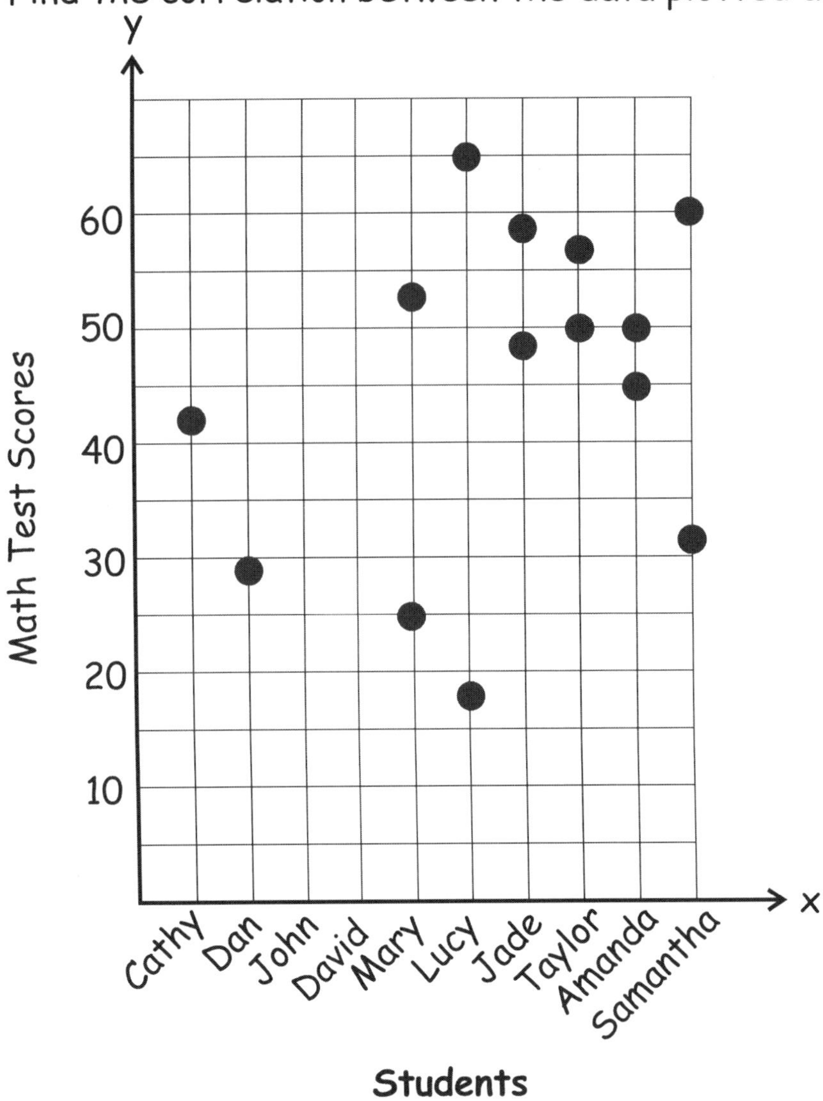

(A) Positive correlation

(B) Negative correlation

(C) No correlation

(D) None

1. Which of the below options matches to the data given ?

7.5 , 10 , 4 , 7.5 , 10 , 6.5 , 6.5 , 9.9 , 9.5 , 9 , 7.7 , 4.5

(A)

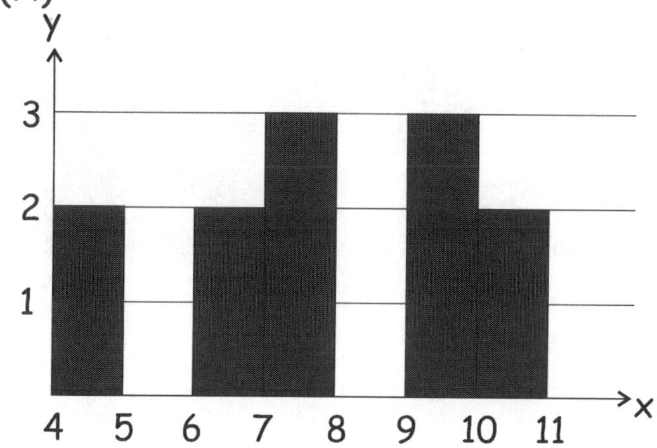

(B)

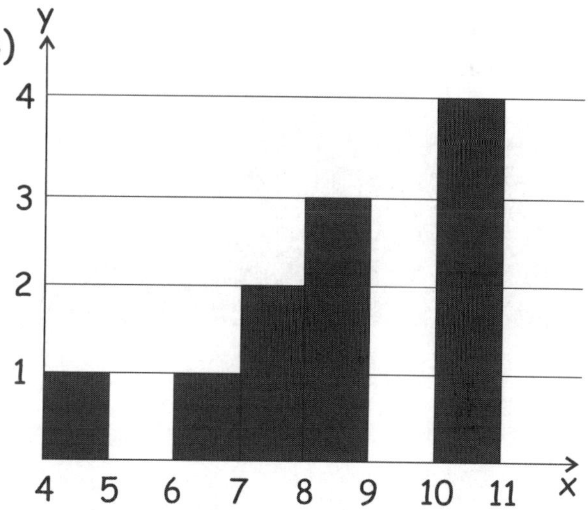

(C)

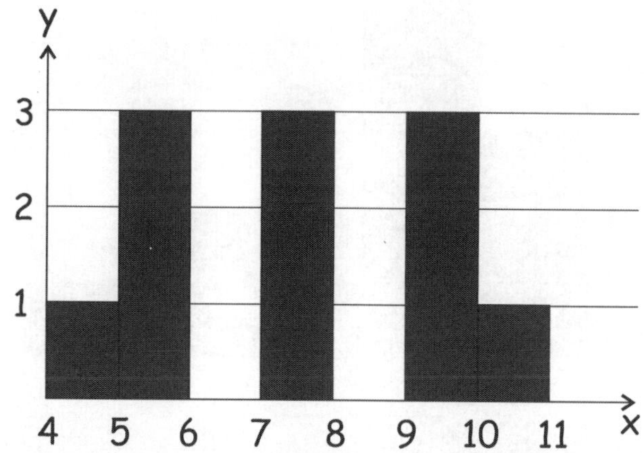

(D)

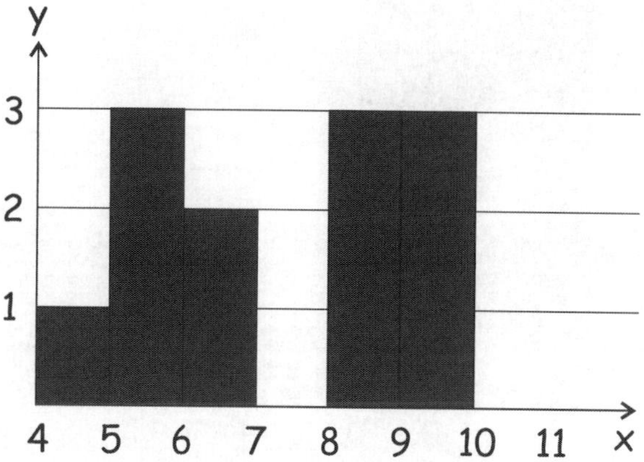

2. Which of the below options matches to the data given ?

2 , 3 , 1 , 1 , 2 , 2 , 2 , 4 , 5 , 2 , 1 , 4

(A)

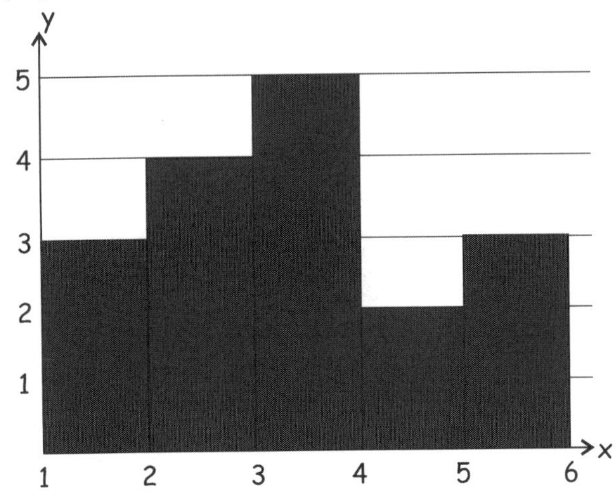

(B)

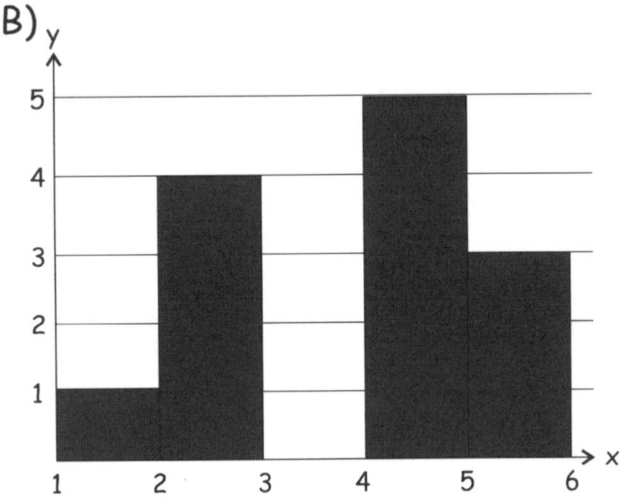

(C)

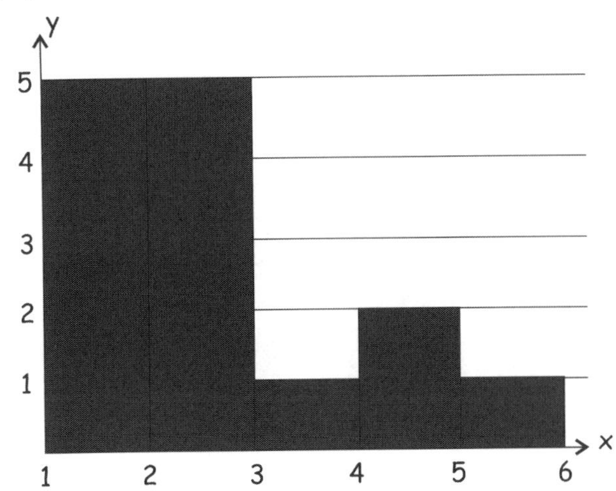

(D)

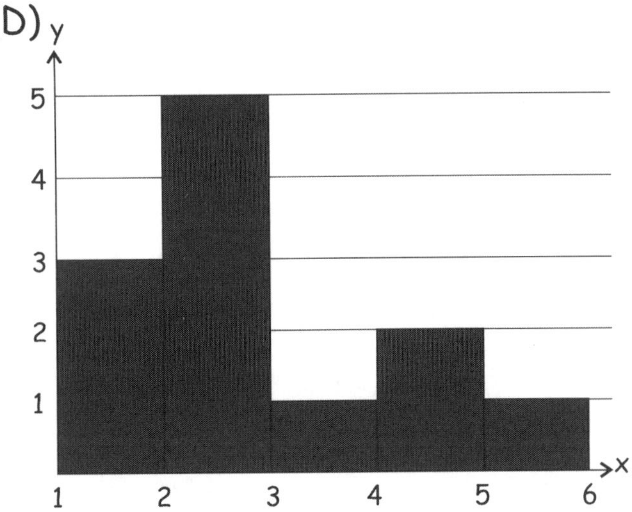

3. Which of the below options matches to the data given ?

5 , 17 , 19 , 23 , 22 , 21 , 7 , 11 , 16 , 18 , 14 , 9 , 16 , 23 , 29

(A)

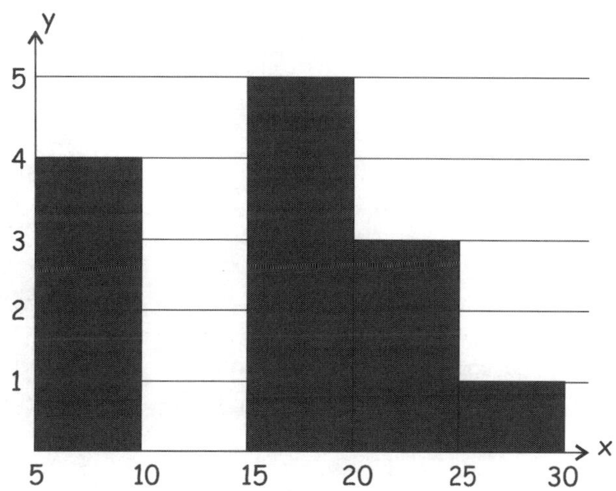

(B)

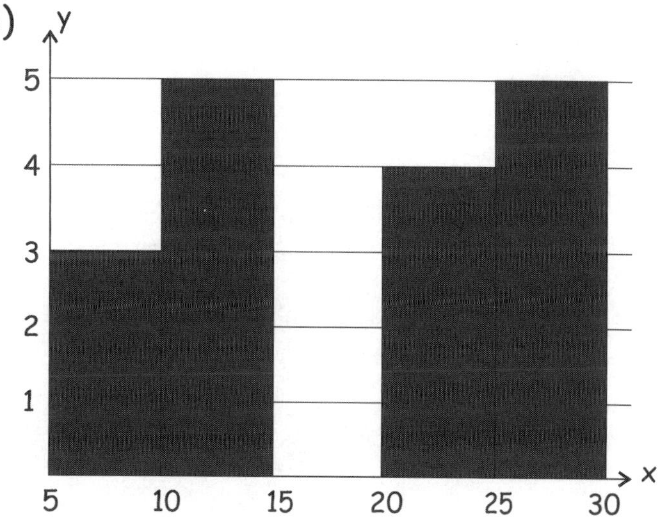

(C)

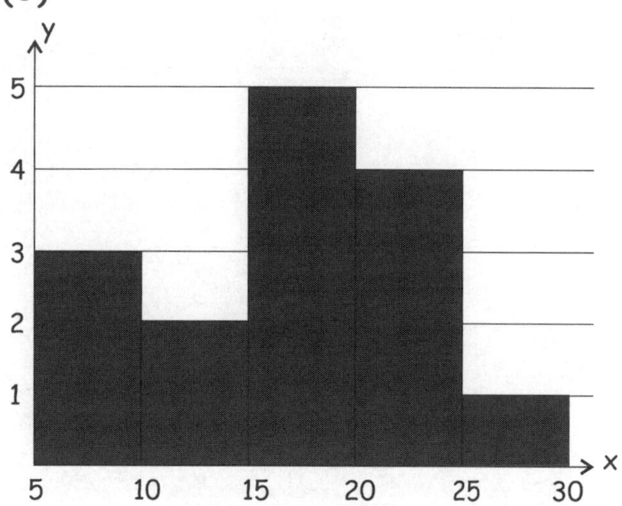

(D)

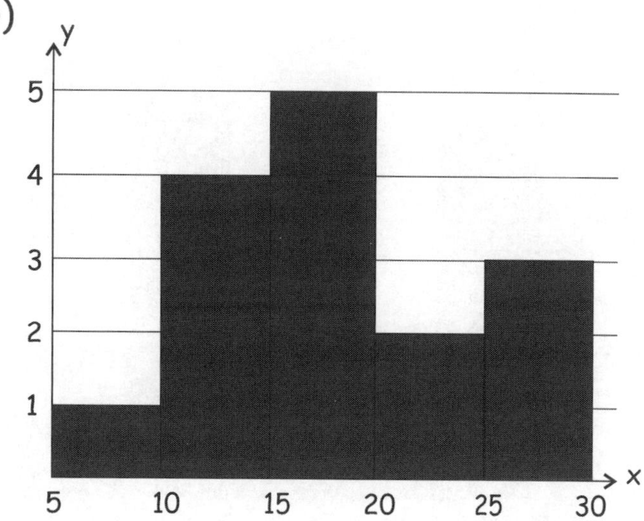

4. Which of the below options matches to the data given ?

6.70 , 6.73 , 7 , 5.70 , 5.65 , 6.70 , 6.72 , 7 , 8.24 , 6.3 , 7.89

(A)

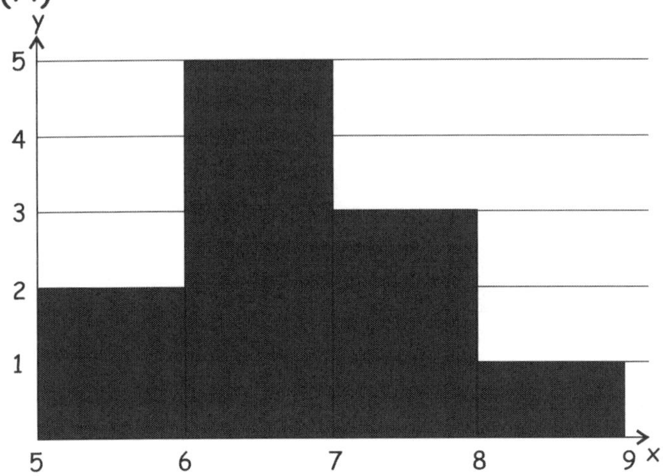

(B)

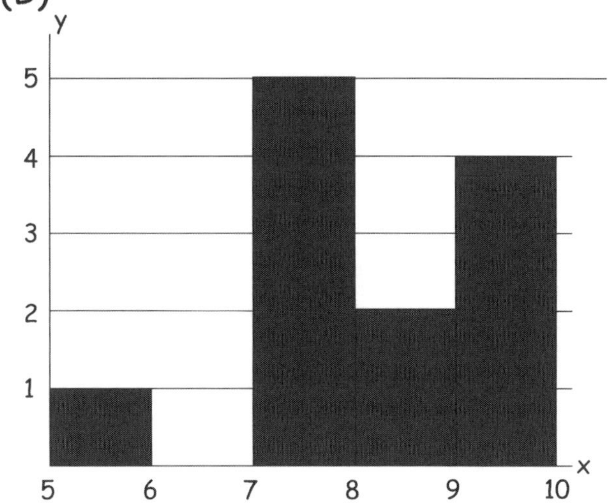

(C)

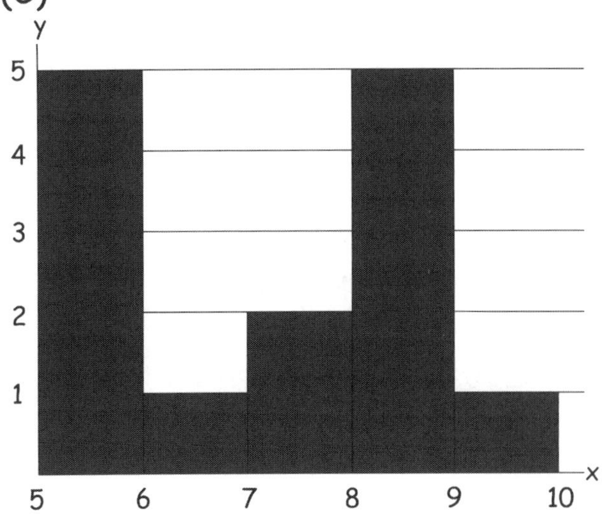

(D)

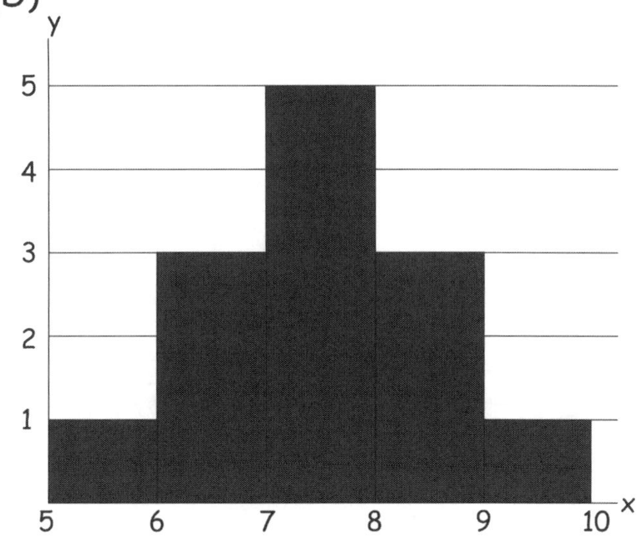

Girls Scout team raised funds through various lemonade stands across the community. Number of lemonade glasses sold are plotted as below. Answer the questions 1 - 4.

Stem	Leaf
0	1 1 1 3 4 4 8 9 9
1	1 5 8
2	6 9
3	
4	8

0 | 1 = 1 Glass

1. Find the median of the above data ?

(A) 6 (B) 7

(C) 9 (D) 5

2. Find the Mean of the above data ?

(A) 4.46 (B) 13.57

(C) 19 (D) 12.47

3. Find the Range of the above data ?

(A) 56 (B) 47

(C) 27 (D) 49

4. What is the Mode of the above data ?

(A) 3 (B) 9

(C) 1 (D) 8

Larry sold a number of town homes over the summer. The house prices are listed below. Based on this data answer the questions 5 - 8.

Stem	Leaf
20	1 8 9
20	9
22	1 2 5
23	4 4 8
24	0 1 9
25	3 7

20 | 1 = $201 K
$201,000

5. Find the median of the above data ?

(A) 209K (B) 225K

(C) 238K (D) 234K

6. Find the Mean of the above data ?

(A) 229.03 K (B) 229.4 K

(C) 228.53 K (D) 227.93 K

7. Find the Range of the above data ?

(A) 67 K (B) 56 K

(C) 54 K (D) 49 K

8. What is the Mode of the above data ?

(A) 234 K (B) 225 K

(C) 233 K (D) 249 K

Grade 6 students took a science quiz and competed for the national science bowl competition. The quiz scores are plotted as below. Based on this data answer this questions 9 - 12.

Stem	Leaf
5	4 9
6	0 8
7	0 1 2 3 4 5 5 5 6 8 9
8	1

5 | 1 = 51 Points

9. Find the median of the above data ?

(A) 73 (B) 73.5

(C) 72.5 (D) 68

10. Find the Mean of the above data ?

(A) 51.44 (B) 77.12

(C) 71.25 (D) 75.21

11. Find the Range of the above data ?

(A) 27 (B) 39

(C) 24 (D) 21

12. What is the Mode of the above data ?

(A) 70 (B) 68

(C) 80 (D) 75

A car garage has repaired a number of cars this Quarter. Based on this data answer the questions 13 - 16.

Stem	Leaf
0	2 2 3 3 3 4 4 4 5 5 6 6 7
1	3
2	
3	8

0 | 2 = 2 Cars

13. Find the median of the above data ?

(A) 11 (B) 4

(C) 2 (D) 7

14. Find the Mean of the above data ?

(A) 9 (B) 11

(C) 7 (D) 5

15. Find the Range of the above data ?

(A) 36 (B) 25

(C) 34 (D) 28

16. What is the Mode of the above data ?

(A) 3 and 4 (B) 7 and 3

(C) 1 and 3 (D) 6 and 4

Grade 5 students took math test and the scores are listed as below. Answer the questions 17 - 20

Stem	Leaf
5	0 0 1
6	1 6 8 9
7	4 4 5 5 5 5 6 6 7
8	1

5 | 0 = 50 Points

17. Find the median of the above data ?

(A) 64 (B) 75

(C) 82 (D) 74

18. Find the Mean of the above data ?

(A) 69 (B) 68.07

(C) 69.69 (D) 70.19

19. Find the Range of the above data ?

(A) 36 (B) 25

(C) 31 (D) 29

20. What is the Mode of the above data ?

(A) 77 (B) 75

(C) 74 (D) 50

Super Soccer team played various games across the fall season.
The number of goals made are plotted below. Answer the questions 1 - 4.

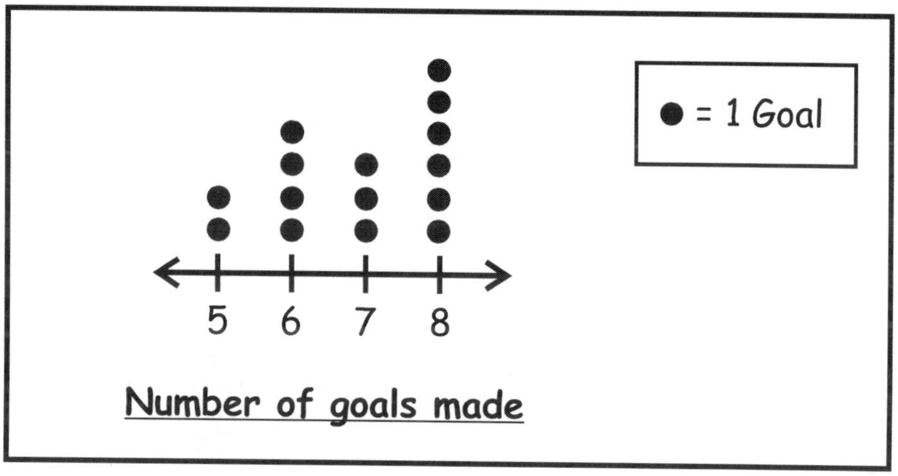

1. Find the median of the above data ?

(A) 7 (B) 8

(C) 11 (D) 5

2. Find the Mean of the above data ?

(A) 6.07 (B) 6.08

(C) 6.05 (D) 6.87

3. Find the Range of the above data ?

(A) 14 (B) 5

(C) 3 (D) 6

4. What is the Mode of the above data ?

(A) 8 (B) 5

(C) 6 (D) 7

Store A sold a number of cars int he last 11 days as shown below. Based on the data answer the questions 5 - 8.

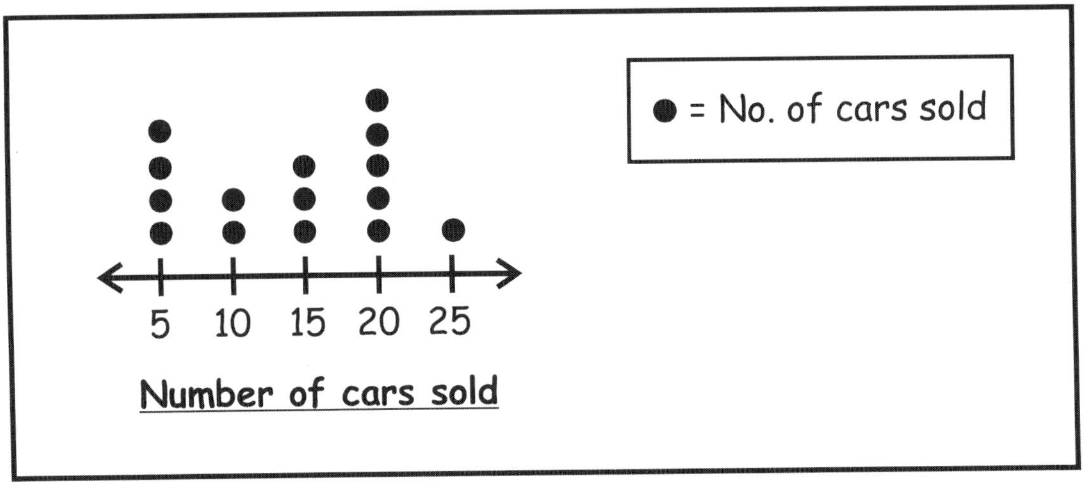

5. Find the median of the above data ?

(A) 20 (B) 15

(C) 10 (D) 17.5

6. Find the Mean of the above data ?

(A) 20 (B) 15

(C) 9 (D) 14

7. Find the Range of the above data ?

(A) 20 (B) 25

(C) 24 (D) 21

8. What is the Mode of the above data ?

(A) 23 (B) 25

(C) 20 (D) 10

Below data gives the average snow fall in Chicago during January 2019. Answer the questions 9 - 12

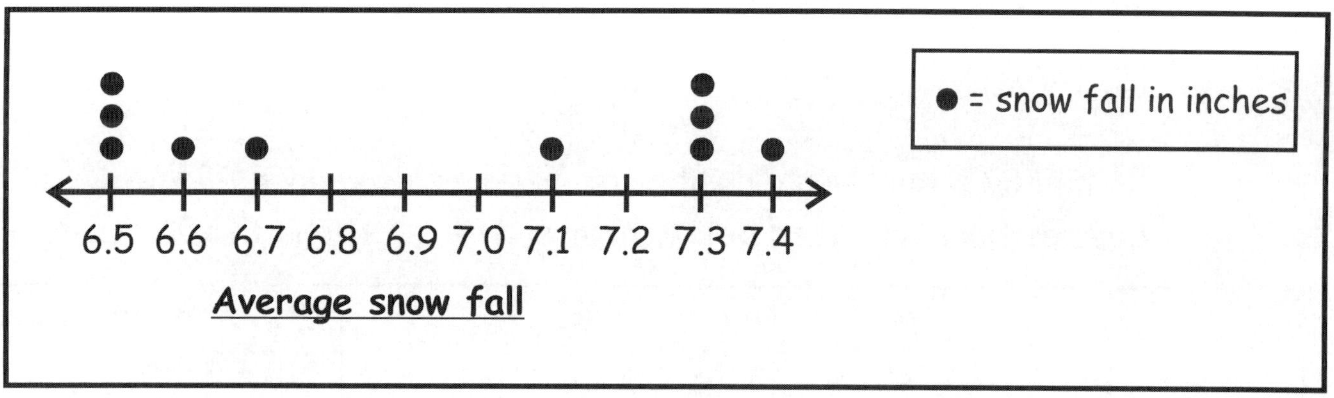

9. Find the median of the above data ?

(A) 6.5 (B) 6.7

(C) 7.3 (D) 6.9

10. Find the Mean of the above data ?

(A) 6.94 (B) 6.34

(C) 6.40 (D) 6.92

11. Find the Range of the above data ?

(A) 1.2 (B) 0.9

(C) 1.5 (D) 0.7

12. What is the Mode of the above data ?

(A) 6.5 and 7.3 (B) 6.0

(C) 6.1 (D) 7.5

RC middle school won a number of medals this year in STEM Competitions a plotted below. Answer the questions 13 - 16

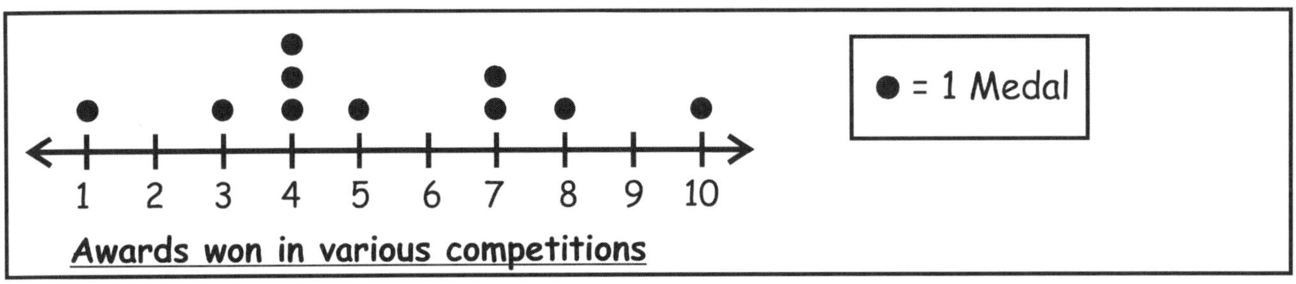

13. Find the median of the above data ?

(A) 3.95 (B) 2.9

(C) 4.5 (D) 4.9

14. Find the Mean of the above data ?

(A) 5.9 (B) 6.53

(C) 5.3 (D) 7.03

15. Find the Range of the above data ?

(A) 5 (B) 7

(C) 10 (D) 9

16. What is the Mode of the above data ?

(A) 4 (B) 9

(C) 7 (D) 2

BR High school conducted a survey on 1800 students for their favorite hobbies. The below pie chart represents percentage of student votes. Answer the questions 1 - 6.

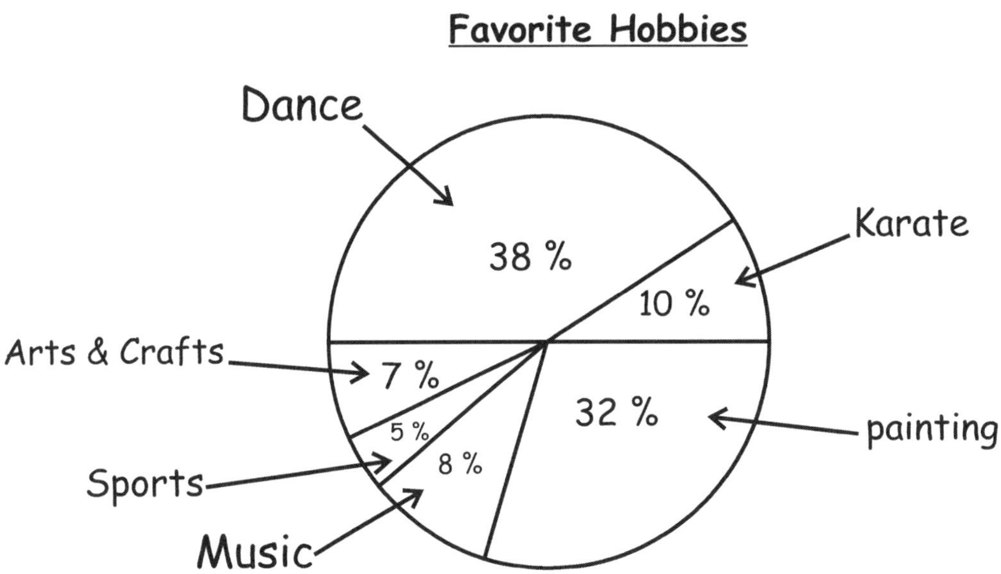

1. Find the number of students who voted for dancing

(A) 648 (B) 684

(C) 864 (D) 664

2. Find the ratio between the number of student votes for painting and karate.

(A) 3:7 (B) 5:16

(C) 16:5 (D) 5:7

3. What percentage of students voted for music than sports ?

(A) 6% (B) 5%

(C) 4% (D) 3%

4. How many students voted for Arts & crafts and painting together ?

(A) 702 (B) 602

(C) 660 (D) 645

5. By what percent is the number of students who voted for Arts & crafts and dancing together are more than those who voted for painting and music together ?

(A) 14.5% (B) 5%

(C) 9% (D) 8%

6. Find the ratio between the number of students who voted for dancing and karate than who voted for painting ?

(A) 2:3 (B) 1:3

(C) 3:1 (D) 3:2

SS High school conducted a survey on 6800 students for their favorite Soft Drinks. The below pie chart represents percentage of student votes. Answer the questions 7 - 12.

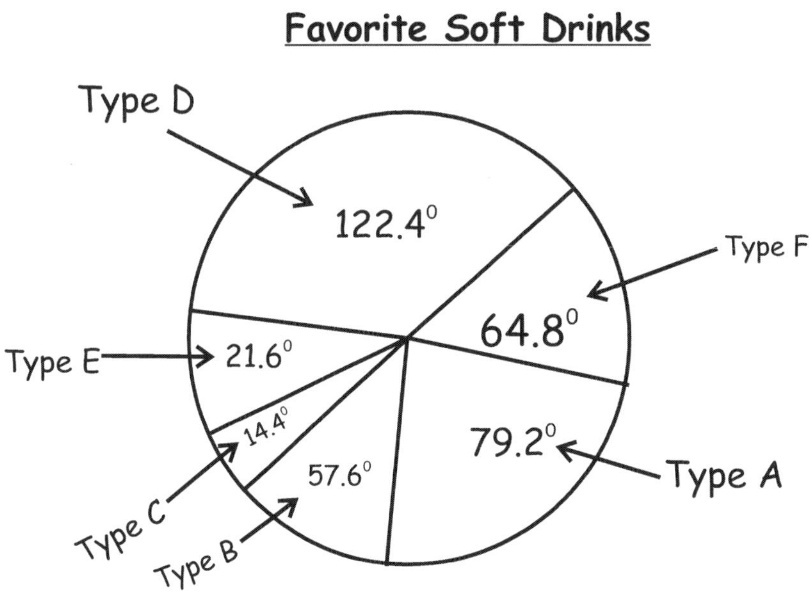

7. What is the difference between the total number of votes for types A and C together and who preferred types D and F together ?

(A) 959 (B) 960

(C) 952 (D) 953

8. Find the ratio of the number of student votes for type F and type A soft drinks

(A) 9:11 (B) 3:13

(C) 6:11 (D) 5:11

9. Find the percentage of the votes received together for types E and F

(A) 18% (B) 20%

(C) 26% (D) 24%

10. How many more votes were received for type C when compared to type D ?

(A) 2754 (B) 2040

(C) 2018 (D) 2900

11. How many votes were received for types B and E together ?

(A) 1496 (B) 1420

(C) 1318 (D) 1485

12. How many votes were received for types B and C together ?

(A) 1290 (B) 1366

(C) 1360 (D) 1450

A happy living County spends the tax amounts received for various community development needs. The below pie chart represents the distribution of funds. Answer the questions 13 - 25.

County funds Allocation

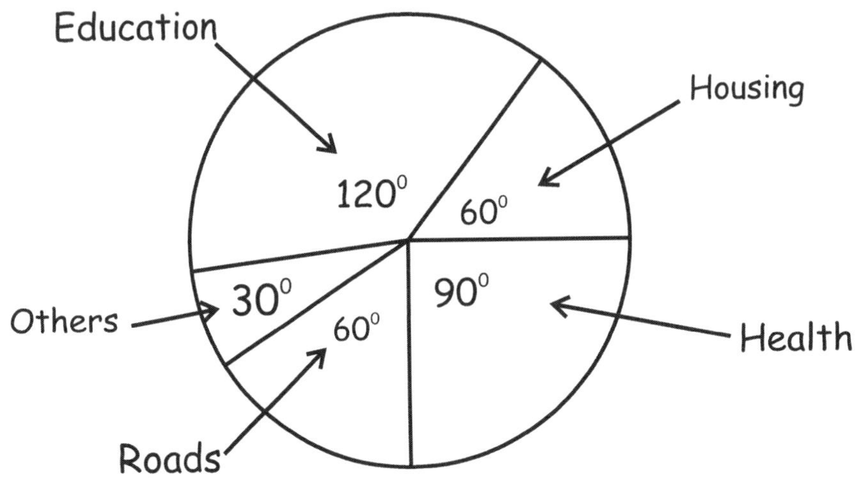

13. Find the ratio of the expenditure on education to health ?

(A) 3:2 (B) 2:3

(C) 3:4 (D) 4:3

14. Find the percentage of money was spent on housing ?

(A) $19\frac{2}{3}\%$ (B) $16\frac{1}{3}\%$

(C) 15% (D) $16\frac{2}{3}\%$

15. The maximum tax money was spent on which of the below ?

(A) Health (B) Education

(C) Roads (D) Housing

16. Which two had the same amount of expenditure from the below ?

(A) Housing and Education (B) Health and Housing

(C) Roads and Housing (D) Housing and others

17. What percentage of money was spent on Education ?

(A) $33\frac{1}{3}\%$ (B) 34%

(C) 32% (D) $33\frac{2}{3}\%$

18. Find the ratio of the expenditure on housing to roads ?

(A) 1:2 (B) 2:1

(C) 3:1 (D) 1:1

19. What percentage of money was spent on roads ?

(A) $12\frac{1}{3}\%$ (B) $13\frac{2}{3}\%$

(C) $16\frac{1}{3}\%$ (D) $16\frac{2}{3}\%$

20. What percentage of money was spent on health ?

(A) 25% (B) 50%

(C) 29% (D) 20%

21. What percentage of money was spent on other items ?

(A) $8\frac{2}{3}$ % (B) 8%

(C) $7\frac{1}{3}$ % (D) $8\frac{1}{3}$ %

22. Find the ratio of the expenditure on others to roads ?

(A) 1:1 (B) 2:2

(C) 1:2 (D) 2:1

23. Find the ratio of the expenditure on education to roads ?

(A) 2:1 (B) 3:2

(C) 2:5 (D) 1:2

24. Find the ratio of the expenditure on health to housing ?

(A) 3:4 (B) 1:2

(C) 2:3 (D) 3:2

25. Find the ratio of the expenditure on education to others ?

(A) 4:3 (B) 4:1

(C) 1:5 (D) 5:1

Zoom Zoom automobiles company manufactures vehicles. The below pie chart represents the production of various types of vehicles. Answer the questions 26 - 35

County funds Allocation

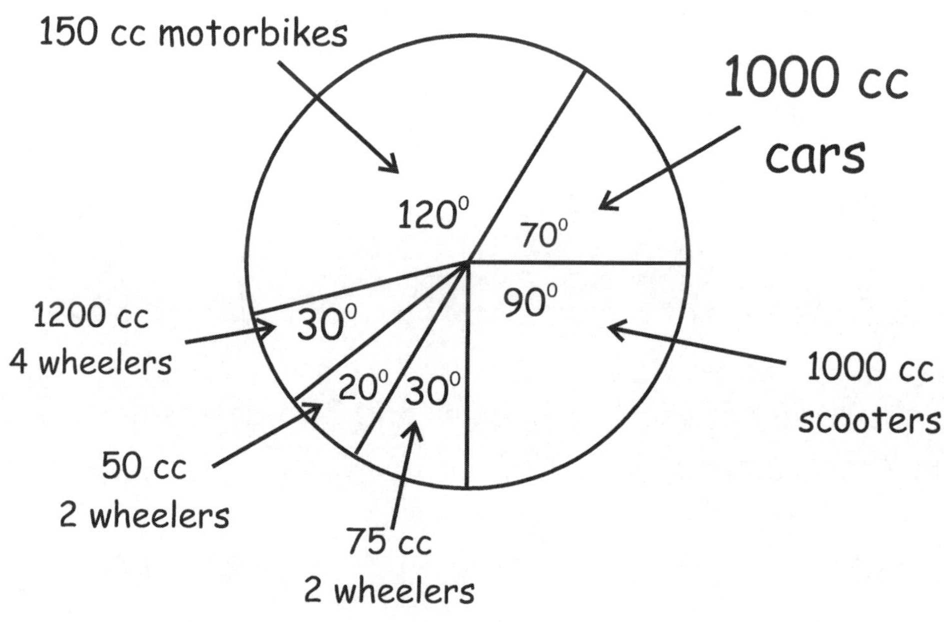

26. Find the ratio between 75 cc 2 wheelers and 50 cc 2 wheelers

(A) 2:1 (B) 2:3

(C) 3:2 (D) 1:3

27. Find the percentage of the production for 150 cc motorbikes ?

(A) 30%
(B) $33\frac{1}{3}\%$
(C) $32\frac{2}{3}\%$
(D) 33%

28. A total of 2700 , 75cc 2 wheelers are manufactured this month. How many vehicles are manufactured in total this month ?

(A) 32,400
(B) 30,840
(C) 32,960
(D) 35,450

29. A total of 7,200 are manufactured this month. How many 150 cc bikes are manufactured in total this month ?

(A) 2,200
(B) 2,400
(C) 2,500
(D) 2,300

30. Find the ratio between 4 wheelers and 2 wheelers produced by Zoom Zoom company ?

(A) 13:5
(B) 5:4
(C) 4:5
(D) 5:13

31. Find the ratio between 150 cc motor bikes and 1000 cc scooters produced by Zoom Zoom company ?

(A) 4:3 (B) 7:4

(C) 6:11 (D) 9:1

32. Find the ratio between 1000 cc scooters and 1000 cc cars produced by Zoom Zoom company ?

(A) 9:5 (B) 7:9

(C) 9:7 (D) 5:9

33. Find the ratio between 1000 cc cars and 75 cc 2 wheelers produced by Zoom Zoom company ?

(A) 3:7 (B) 7:4

(C) 1:7 (D) 7:3

34. Find the ratio between 1200 cc 4 wheelers and 75 cc 2 wheelers produced by Zoom Zoom company ?

(A) 1:1 (B) 1:4

(C) 8:1 (D) 2:1

35. What percentage of 1,200 cc 4 wheelers were manufactured in the Zoom Zoom company ?

(A) $6\frac{2}{3}$ % (B) $9\frac{2}{3}$ %

(C) $5\frac{1}{3}$ % (D) $8\frac{1}{3}$ %

36. If the company produced 3,600 vehicles in total, how many 50 cc 2 wheelers are produced ?

(A) 320 (B) 200

(C) 290 (D) 274

IAAT Vol 1
Work Book
Answer Keys

	#1	#2	#3	#4
Q #	Fractions Multiply	Fractions Divide	Fractions Subtractions	Decimals Add_Subtractions
1	A	C	B	B
2	A	B	B	C
3	D	C	C	A
4	A	D	A	A
5	A	C	A	D
6	B	C	D	D
7	B	C	C	D
8	D	B	B	C
9	C	C	D	D
10	B	C	B	C
11	A	C	B	C
12	A	A	B	A
13	B	D	B	C
14	C	D	D	D
15	A	D	B	C

Q #	#5 Decimal Multiplication	#6 Decimal Divisions	#7 Integer Multiplication	#8 Scientific Notation
1	A	B	B	B
2	A	D	A	C
3	D	C	B	B
4	C	A	C	B
5	C	C	C	D
6	D	C	A	B
7	A	B	C	B
8	A	A	C	B
9	A	B	D	A
10	A	C	C	D
11	B	B	D	A
12	B	D	D	C
13	D	C	D	B
14	A	B	A	C
15	C	A	D	C

	#9	#10	#11	#12
Q #	GCF	GCF Monomials	LCM Numbers	LCM Monomials
1	C	C	D	C
2	B	A	A	C
3	A	A	D	B
4	D	D	A	C
5	C	B	B	C
6	B	D	C	A
7	B	D	D	B
8	B	D	B	A
9	A	A	C	C
10	B	B	D	C
11	C	A	D	A
12	D	C	A	C
13	C	C	D	D
14	B	C	C	B
15	B	D	B	B

	#13	#14	#15	#16
Q #	Order of Operations	Verbal Expressions	Verbal Expression Equation	Monomials
1	A	C	A	B
2	A	B	B	C
3	D	C	D	A
4	D	D	A	C
5	D	B	C	C
6	B	B	D	D
7	B	C	A	C
8	B	B	B	C
9	D	D	D	B
10	C	A	B	B
11	B	C	D	A
12	A	B	C	B
13	C	B	B	A
14	A	A	C	A
15	B	C	A	D

	#17	#18	#19	#20
Q #	Inequalities	Evaluate Expressions	Evaluate Expressions	Solve for X
1	C	C	B	C
2	A	A	D	D
3	C	A	C	C
4	D	B	B	B
5	B	A	B	C
6	B	B	C	B
7	C	B	A	B
8	D	D	C	C
9	D	B	B	B
10	B	A	C	B
11	D	D	B	C
12	A	D	B	C
13	D	A	C	C
14	C	D	A	A
15	C	D	A	C

	#21	#22	#23	#24
Q #	Absolute Value	Absolute Value	Proportions	Proportions
1	C	A	B	C
2	A	B	C	A
3	C	C	C	C
4	A	D	C	A
5	B	A	A	C
6	A	D	C	D
7	D	A	B	A
8	A	B	A	A
9	A	A	A	B
10	B	B	A	D
11	B	D	A	C
12	D	B	B	C
13	A	C	A	B
14	D	B	A	B
15	B	D	C	D

	#25	#26	#27	#28
Q #	Percent Discount	Percent MarkUp	Percent Tax	Percent Change
1	D	C	C	D
2	B	D	A	B
3	D	D	A	C
4	D	B	B	B
5	A	C	D	D
6	A	A	A	D
7	B	B	C	C
8	C	B	A	C
9	D	C	C	A
10	A	D	B	A
11	A	D	A	B
12	C	D	D	C
13	D	A	C	A
14	A	B	C	A
15	B	B	C	B

Q #	#29 Radicals	#30 Radicals	#31 Radicals	#32 Radicals
1	B	B	D	D
2	A	D	B	C
3	B	B	D	A
4	D	A	C	D
5	B	D	D	D
6	A	A	B	C
7	B	C	D	B
8	C	D	C	D
9	B	B	D	A
10	D	D	D	B
11	C	C	B	C
12	B	D	B	A
13	B	C	A	C
14	D	B	A	D
15	B	A	A	A

	#33	#34	#35	#36
Q #	Radicals	Radicals	Triangles	Rectangle-Square
1	A	B	C	D
2	A	C	C	C
3	D	D	A	B
4	B	A	D	D
5	A	A	A	B
6	C	A	C	C
7	D	C	B	D
8	B	B	C	B
9	C	D	B	A
10	A	B	B	D
11	B	C	C	D
12	D	D	D	C
13	C	D	D	B
14	C	C	B	D
15	A	A	D	A

	#37	#38	#39	#40
Q #	Parallelogram Trapezium	Circle - Area - Circumference	Volume-Sphere	Volume-Rectangle Square prisms
1	C	D	A	A
2	C	A	D	D
3	D	D	B	B
4	A	B	A	D
5	C	A	C	B
6	A	C	D	D
7	A	C	B	A
8	D	C	D	A
9	A	B	A	B
10	A	A	C	D
11	C	A	A	C
12	C	D	B	A
13	A	A	C	C
14	D	D	B	D
15	A	C	C	D

	#41	#42	#43	#44
Q #	Volume - Cone - Cylinder	Additive Inverse	Multiplicative Inverse	Rational - Irrational Numbers
1	B	D	A	A
2	A	A	C	D
3	C	C	C	A
4	C	C	B	C
5	B	B	A	B
6	A	D	D	D
7	C	A	B	B
8	A	A	C	D
9	A	C	A	A
10	B	B	D	C
11	C	A	C	B
12	A	D	C	D
13	C	C	B	A
14	B	B	D	C
15	C	D	C	B

	#45	#46	#47	#48
Q #	Exponents	Slope- 2 points	Slope intercept	Slope Graph
1	C	A	B	D
2	A	B	C	B
3	A	B	B	B
4	D	C	A	D
5	B	A	D	A
6	A	C	C	D
7	C	D	C	B
8	C	D	D	A
9	D	D	A	B
10	D	B	D	A
11	A	C	B	A
12	D	C	A	B
13	C	A	A	B
14	B	C	A	D
15	B	A	A	A

	#49	#50	#51	#52
Q #	Straight line slope	Parallel line slope	Perpendicular line slope	Missing angle
1	D	C	A	C
2	A	A	B	D
3	A	A	B	C
4	D	C	D	C
5	A	C	B	D
6	D	B	C	A
7	C	C	A	B
8	B	A	C	A
9	C	D	B	C
10	A	A	A	C
11	B	B	A	D
12	B	A	D	A
13	D	C	A	B
14	A	B	D	B
15	C	C	B	D

	#53	#54	#55	#56
Q #	Missing angle	Reflection	Rotation	Translation
1	B	A	C	B
2	D	D	B	C
3	D	D	A	B
4	A	B	D	B
5	A	D	B	A
6	A	A	C	A
7	C	B	B	D
8	B	C	C	C
9	A	B	D	C
10	C	C	B	D
11	A	B	B	B
12	B	A	B	B
13	C	A	D	C
14	B	D	A	A
15	C	A	B	D

	#57	#58	#59
Q #	One step word problems	System of equations	System of equations
1	A	B	A
2	D	A	D
3	A	A	A
4	A	C	A
5	D	C	C
6	B	D	A
7	B	D	B
8	A	D	D
9	D	C	D
10	A	C	A
11	B	D	A
12	B	A	C
13	C	A	A
14	B	D	B
15	A	C	A

	#60
Q #	System of equations
1	A
2	D
3	D
4	D
5	A
6	C
7	D
8	C
9	D
10	B
11	D
12	A
13	C
14	B
15	B

#61 Line graph - Answer keys

1. D
2. A
3. C
4. C
5. B
6. D
7. A
8. C
9. A
10. D
11. B
12. A
13. B
14. C
15. A
16. D
17. A
18. D
19. C
20. B

#62 Scatter plot - Answer keys

1. C
2. B
3. A
4. A
5. C

#63 Histogram plot - Answer keys

1. A
2. D
3. C
4. A

#64 Stem leaf plot - Answer keys

1. C
2. D
3. B
4. C
5. D
6. B
7. B
8. A
9. B
10. C
11. A
12. D
13. B
14. C
15. A
16. A
17. D
18. A
19. C
20. B

#65 Dot plot - Answer keys

1. A
2. D
3. C
4. A
5. B
6. D
7. A
8. C
9. D
10. D
11. B
12. A
13. C
14. C
15. D
16. A

#66 Pie chart - Answer keys

1. B
2. C
3. D
4. A
5. B
6. D
7. C
8. A
9. D
10. B
11. A
12. C
13. D
14. D
15. B
16. C
17. A
18. D
19. D
20. A
21. D
22. C
23. A
24. D
25. B
26. C
27. B
28. A
29. B
30. D
31. A
32. C
33. D
34. A
35. D
36. B

Made in the USA
Middletown, DE
27 August 2024

59739347R00203